T0224517

KAIST Research Series

Series editor

Si Kyoung Roh, Daejeon, Korea, Republic of South Korea

For further volumes:
http://www.springer.com/series/11753

Kwang Hyung Lee

Three Dimensional Creativity

Three Navigations to Extend our Thoughts

 Springer

Kwang Hyung Lee
Bio and Brain Engineering
KAIST
Daejeon
Korea, Republic of South Korea

ISSN 2214-2541   ISSN 2214-255X   (electronic)
ISBN 978-94-024-0783-9   ISBN 978-94-017-8804-5   (eBook)
DOI 10.1007/978-94-017-8804-5
Springer Dordrecht Heidelberg New York London

Printed on acid-free paper

Springer is part of Springer Science+Business Media (www.springer.com)

Preface

What is creativity? Simply put, it is the power to create a new idea. A new idea means that you are different from another person or the man you were yesterday. Everyone says that creativity has gotten more important since the society has become complicated and uncertainty has increased. As the society and technology change rapidly, any knowledge, no matter how important and good, may become of no use in 10 years.

We may ask ourselves: How much effort and time am I putting into developing that important creativity? Most would answer not even more than an hour. It is strange, is it not? Even though we know the importance of developing creativity, we put no effort into it at all. No class of creativity is conducted at school. No training course for creativity is offered at companies.

Have You Invested at Least One Hour to Develop Your Creativity?

Why don't people invest time in cultivating their creativity? It may be that people think creativity is not something they can earn by sowing effort. They may think that creativity is something that people are born with, and it cannot be enhanced even if they work at it in a classroom.

Is creativity something that we are simply born with and cannot hone with effort?

Some say that reading a lot of books contributes to developing creativity. Some say that setting aside some time to imagine a lot works. Some say that if you focus on a certain matter, a new idea may come up. Indeed, such methods can contribute to enhancing creativity, but they are too vague. It is similar to saying you have to "practice very hard" to be good at playing soccer. The fact that you hear only those vague words indicates that people do not fully understand what creativity really is.

In general, our thinking sticks to reality unless there is any external stimulation. Thinking is fixed on a point of time, a position, and an area that we are interested in. With our thinking fixed on a certain situation, it becomes difficult to come up with a fresh idea. A fresh idea is likely to come up when our thinking is freed. When it is absorbed with an object, we cannot say that it is freed. We may see those around us who imagine a lot to create many original ideas. They make countless new ideas since their thinking is unhampered.

Questions Hinder Your Mind from Sticking to Reality

To come up with new ideas, it is necessary to break your adherence from reality. Once you do this a new environment will be ahead of you and your brain will be stimulated by it. The stirred brain is activated to create new thoughts. That is why some say that traveling or reading books helps create new ideas. These activities spur the individual, directly or indirectly, toward a new environment.

It can be challenging, however, for us to go on a trip or read books constantly. Is there any other way to stimulate the brain?

Yes, there is, and it is by "questioning." The brain is kindled when given a question as the brain concentrates on what the question presents. When a question arises on the Winter Olympics in 2026, the thinking is directed to the year 2026 and to winter sports.

When a new problem comes up at school or work, people may gather together for brainstorming as a way of solving the problem. Brainstorming is a discussion among people in which they question each other. What they say becomes a question to another, and what others say becomes a question to them. While exchanging words, they may come up with new ideas, and often they are directed to the solution to solve the problem. In other words, asking questions rouses the brain and makes it produce fresh ideas.

Although people may not question you, you may ask yourself questions to constantly stimulate your brain even when you are alone. New ideas are churned if you continuously throw questions at yourself. This routine gradually will help you become a creative person.

Two issues are involved in developing creativity: First, what questions would you ask yourself? Second, how can you make yourself ask questions?

What Questions Would You Ask Yourself?

Let us think about the first point. A question needs to be universal so that it can be applied to various areas. The question has to be about the basic elements of things. In this regard, three questions are suggested as follows:

(1) Question of Time: When a problem arises, question yourself on a time axis first of all. How would the matter be handled 10 or 20 years from now? Such question will stimulate the brain, and this line of thought will move to the point of time that the questions lead to. When directed to a new point in time, the brain starts to imagine the environment in that point in time.
(2) Question of Space: When given a problem, ask yourself some questions on its spatial elements. How would the matter be handled in Saudi Arabia? How would the matter be perceived in China? How would the shape change? Think of the given problem in another angle of space. In the process, you may encounter a new environment, which will stimulate your brain and make it come up with a fresh idea.
(3) Question of Field: Apply the given problem in a variety of fields. How would the matter be handled in the field of music? How would it be treated in the field of electronic engineering? If you think of the matter in different fields, a convergence among fields takes place, and new ideas are likely to emerge.

Because the three elements explained earlier are basic components consisting of almost everything in the world, everything can be defined by them. Questions about them can touch almost all the important aspects of given problems.

Asking yourself the three questions above will make your brain travel in a new environment more freely. True, this is a theory. A theory, no matter how good it is, would be of no use unless you apply it. For instance, no matter how good you are at reading musical scores, you cannot play music well unless you practice playing frequently and get familiar with an instrument. You need to repeatedly practice to the point of making something your habit.

How Could You Make Yourself Ask Questions?

All mental behaviors of humans originate from the brain. All its processes are recorded in its circuit that consists of brain cells. It is a neural circuit that controls recognition and judgment. Memorizing new facts is a process that creates a neural circuit. Playing repeatedly to be familiar with music and adjusting to local time in an overseas trip involve the formation of a neural circuit. Every memory and every habit are made by a neural circuit, and such a neural circuit is never formed at once; rather, repeated efforts are necessary to make one. This is why you need repetition to memorize a new word or be adept in playing music.

What makes an individual repeat a pattern? Humans resort to pleasure. Pleasures are a strong element that induce human behavior. Pleasures depend on dopamine, a neural substance released from the brain. Humans secrete dopamine when commended. Commendation pleases a person and makes him want to be commended again. Such repetition results in the formation of a neural circuit. Commendation is important when it comes to repeated efforts. It leads to repetition, forming a neural circuit, a habit, and a good disposition in an individual. Parents and teachers need to give commendations to elicit good conduct.

But how could you make yourself ask questions when alone? It is necessary to make it a habit to ask the three questions of time, space, and field in such moments. When it becomes your habit to ask yourself questions, a lot of new ideas will come to you. Make a neural circuit that asks the three questions. Self-questioning is necessary when a problem that needs to be solved is encountered. As the process of questioning is repeated, it will become a habit, and this ingrained tendency will help you to become a creative person.

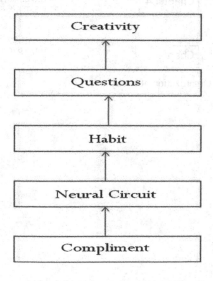

[Content flow of creativity]

The Three-Dimensional Navigation That Expands Your Thinking

The three questions suggested here are given on the three axes: the axis of time, the axis of space, and the axis of field. Combining these three results in forming a three-dimensional world, and you can free yourself from reality as you move on the three axes in the three-dimensional world. In such a free state, the brain produces a wealth of new ideas.

The three questions are universal and applicable to every problem. They are subdivided so that they can be readily applied to practical matters: The question of T (time) is divided into T1 (Transpose), T2 (Tempo), and T3 (Translation). The question of S (Space) is divided into S1 (Shape), S2 (Site), and S3 (Size). The question of F (Field) is divided into F1 (Function), F2 (Fertileness), and F3 (Fusion). These categorized rules are applied on the three axes, respectively.

When a brain is given a question, thoughts move along the direction the question points out. Thus, questions play the role of directing our thinking. This navigation prompts us to explore another area. When navigating another place, we encounter a new environment and tend to produce new ideas. Hence, the three questions suggested here can be spoken of as a navigation that expands our thinking, and the three-dimensional axes a frame that enables us to think three dimensionally.

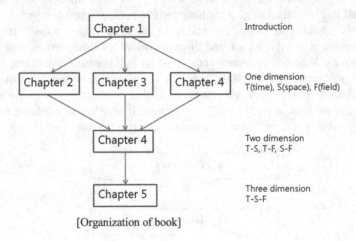

[Organization of book]

Kwang Hyung Lee

Contents

Chapter 1
Creativity Can Be Cultivated

Abstract Can creativity be cultivated? Or is it something inherent that cannot be built despite one's efforts? This chapter gives an answer to this question. New ideas come from stimulation of brain. In general, we are fixed on reality, and thus a new idea doesn't come up easily. Asking questions stimulates the brain to release us from reality, and repeating such questions forms the habit of asking questions that increases creativity. Repeating forms neural circuit corresponding to the habit in the brain. Three types (time, space, field) of questions are recommended.

Keywords Creativity • Asking questions • Habit • Neural circuit • Neural network • Time • Space • Field • Brain

Canoeing and rafting are two popular boating games. Canoeing is an event for paddling along a fixed course on a serene lake, in which it is important to exert power uniformly at the captain's command, and rafting is paddling downward in the valley along the river, in which no one knows when he will be confronted with a rock or a whirlpool. New situations develop all the time, for which there is no fixed manual. Therefore, all members of a team must be on strict alert for problems that may appear at any moment.

Living in modern society defying the forecast of 1 year later is something like rafting. For one thing, it is hard to predict the world's economy satisfactorily for the next month. We are in the dark about whether another Steve Jobs will appear next year after the earthquake in the industrial world. This really convinces us that we are living in the age of uncertainty.

Creativity shines up even more in such times. The future does not unravel in a definite course and is hard to prepare for in advance. We have no choice but to respond anew at every turn of the moment. Under these circumstances, standardized knowledge is outshone by creativity that can devise a new idea to work out a problem. For such reason, it is said that in the twenty-first century, "creativity" is the most important trait we need to be equipped with (Fig 1.1).

K. H. Lee, *Three Dimensional Creativity*, KAIST Research Series,   1
DOI: 10.1007/978-94-017-8804-5_1, © Springer Science+Business Media Dordrecht 2014

Fig.1.1 Rafting and canoeing

Can creativity be cultivated? Or is it something inherent that cannot be built despite one's efforts? Answering this question requires an attempt to define creativity. Creativity is the characteristic of thinking differently from others. It means having different thoughts from the next one or from yesterday as a self. Creativity also means the ability to come up with a new idea.

1.1 Creativity Starts with Asking Questions

Analysis shows that around 20 % of the Nobel Prize winners in history are Jews. Since there are about 16 million Jews living on earth, this number corresponds to only around 0.22 % of the 7 billion population of the world. So what makes it possible for a people accounting for only 0.22 % to take away 20 % of the Nobel

Prizes? Isn't that something enigmatic? Can the Jews be born with special genes related to creativity? It is biologically impossible.

The Jews have their people's identity maintained through their maternal line. If a mother is Jew, her child also becomes a Jew whether the father is Jew or not. Since genes are mixed at random, it is impossible to keep preserving the mother's genes and hand them down to children. Thus, we cannot assume that the Jews are born with special genes for creativity.

I have once met and asked a Jewish friend about their creativity but he said he didn't exactly know the explanation for it. He did say, however, that he had heard a lot since childhood about the virtue of asking many questions, asserting oneself, and discussing matters. He said he had made many discussions at home with his parents and been encouraged at school to ask many questions, adding that Talmud, the traditional textbook for Jewish education, says so, too. I have finally found that it is Talmud education that makes the Jews what they are.

Education means excavating and developing talents, not just looking at what is natural but developing ability by artificial endeavor as well. So is creativity. Some people say that creativity is something natural and cannot be acquired by endeavor, but actually they don't know the mechanism of creativity. As the Jewish Talmud implies, asking many questions and generating discussions can cultivate creativity.

When we are fixed on reality, a new idea doesn't come up easily. In this case, getting out of reality can bring forth a new idea. Since asking questions stimulates the brain to release us from reality, repeating such questions forms the habit of asking many questions that increases creativity. Asking questions frequently requires a third party's praise each time. In short, it is asking questions and obtaining compliments that cultivate creativity. This means positive enforcement is important to building creativity (Fig. 1.2).

Incidentally, the fact that this kind of repeated endeavor can trigger creativity can be fully explained by the theory of neuroscience. What we humans think and remember are brought about by a network of neurons. Synapse among neurons constructs a connection and creates a network among neurons. If you repeat something by endeavor, it builds a neural network and this builds a habit. Practicing a foreign language or playing an instrument also involves building a neural circuit.

1.2 Methodology for Developing Creativity

When we need to emphasize or learn something, it begins with a methodology. For example, in learning to play soccer, there is a way of practicing it. Practicing running, kicking a ball far, passing a ball, shooting, etc. will lead us to being a good soccer player in the long run. Likewise, there is a way of learning to play the guitar. We hold the strings of the guitar and play a few keys that make a chord. After combining these keys and producing continuous sounds, we create music.

However, there seems no way to develop creativity, which everyone says is important. Being asked how to increase creativity, they simply say reading many books or

Fig.1.2 Flow of creativity development

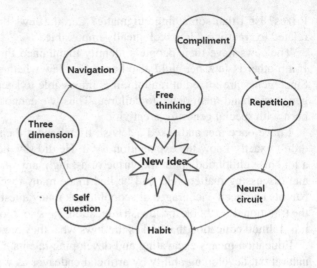

thinking often leads to creativity. It sounds like one should practice much in order to play the soccer well. How much would it serve the one who was asked to play the guitar well if he were just told to practice without any concrete way presented?

This book was written to answer precisely how to develop creativity. The theory for developing creativity referred to in this book chiefly follows theoretical development as described below.

- Creativity is hitting upon a new idea.
- A new idea is likely to occur while asking questions.
- Asking questions to ourselves helps release us from reality.
- A habit can be formed by creating a neural circuit.
- A neural circuit is built through repetition.
- Giving compliment is important for reinforcement of the neural network.
- Building a habit of asking questions increases creativity.
- It is while we are alone that we can nurture the habit of asking.
- Ask questions by extending thoughts along the three axes of time (T), space (S), and field (F).
- Cultivate the habit of asking questions in three-dimensional TSF.
- Taking the habit of asking three-dimensional questions can make a creative person.
- By knowing the mechanism of creativity, we will not simply give up and say that being creative is natural.

1.3 Extend Your Thoughts Guided by the Three-Dimensional Navigation

The writer has long had a tendency to think of what will happen 5 years later or how things will change 10 or 20 years later. It seems that I have paid even more attention to the future due to my major of Computer and Bio-and-brain

Engineering, which are fields that undergo change extremely fast. Not only to myself but also to my students have I mentioned words oriented toward the future or change of "time."

With the advancement of information and communication technology, the concept of distance has changed. It has become possible to exchange information even with people far away as if they were next to us. The world has become something like a small city and to adapt to this kind of change, people have been forced to talk about globalization, or consider pulling down the wall caused by the walls in "space."

As modern society turns even more complex, the field is becoming departmentalized. The complexity of social phenomena is hard to comprehend totally with human ability, so people have come to concentrate on only part of it. This concentration on a particular discipline results in not knowing what is happening in a field next to ours. Thus, looking around many spheres by converging "fields" can lead to a new idea.

The three elements (time, space, field) explained earlier are basic components of almost everything in the world. Almost all the things can be defined by them. When a problem is given, it's important aspects can be touched by questions about the three elements.

While if simply being unchanged we are likely to adhere to reality and be entrapped in a fixed idea, moving along the three axes (time, space, field) mentioned earlier can get us out of that reality. We can escape from a fixed idea along the three navigations. This book is intended to present a three-dimensional navigation (i.e. questions) and explain the way to escape from a fixed idea with this navigation. It is also a method to cultivate creativity by extending our thoughts with this navigation of three components (Fig. 1.3).

1.4 Creativity Made by a Neural Circuit

The three axes of TSF (time, space, and field) are bound together into one frame, that is, three-dimensional creativity. This is the combination and integration of TSF components already existing as individual elements. By combining three dimensions, a three-dimensional world is formed and traveling this three-dimensional world allows escaping from a fixed idea because we cannot be restricted to reality anymore. Like the prince in *Le Petit Prince* by Antoine de Saint-Exupery, we can watch the world from outside of the earth. Escaping from a fixed idea releases our thinking (Fig. 1.4).

What I want to emphasize here is the fact that we must make it a habit to ask questions to ourselves along the TSF navigation. This is just a theory and if we simply follow a theory and remain there, we cannot learn from experience. However enthusiastically we may "read the music," we won't be able to perform music well. Instead, we must keep practicing while reading music until it becomes our habit since it is all repetition that makes a neural circuit.

All our memory, thinking, and action are managed by the brain. In our brain, there are 100 billion neurons. Each neuron, singly, cannot function properly although it

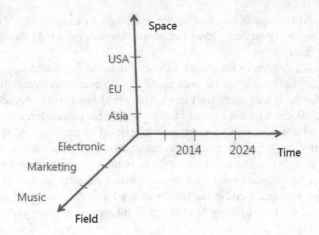

Fig.1.3 Three dimension axes

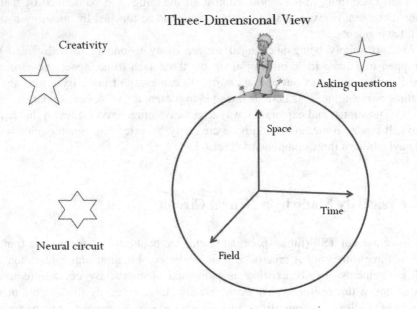

Fig.1.4 Development of three-dimension creativity

has not been clearly identified yet what each of them does. Neurons cannot function smoothly unless they do things jointly in mutual interaction. Synapse connects neurons and acts similarly as in an electronic circuit, where numerous electronic components do not each function independently but do definite things only when they are interconnected to make a circuit (Fig. 1.5). This is similar to the case of a company in which a single worker cannot perform a big duty properly alone. The given responsibility cannot be performed well until many people help one another in mutual cooperation. In short, in the brain, too, neurons cannot perform their own functions

Fig. 1.5 Synapse makes the connection between neurons

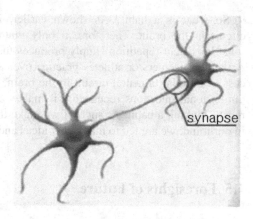

until a circuit is made among them. Here, functions refer to knowing things by heart, rendering a decision, and carrying it into practice.

An electronic circuit, which is an inanimate object, does not change once the circuit has been made. However, since neurons are animate objects, a neural circuit can change along time. If neurons are made to cooperate often, a strong connection forms among them to turn out some sort of teamwork, It is like in a company. If workers are made to cooperate often, teamwork is developed. In the same manner, things easily repeated with neurons through teamwork (memory, judgment, and action) are called a habit.

Moving along in life, we pick up many habits. These habits are nothing but the result of a circuit made though the teamwork of neurons. In a company, too, a department with good teamwork implements an order immediately or carries it out on its own judgment. A department without teamwork fails to perform properly even what it is told to do.

Given a question, we think hard to answer it. If we ourselves were able to ask a question, this would set us thinking voluntarily to answer it. Asking questions to ourselves helps us to have different thoughts releasing us from reality. Asking questions of our own accord can bring forth many new ideas. This is the very basic theory for cultivating creativity and all we have to do is nurture the habit of asking our own questions in our mind.

After which, in order to create a new idea, what kind of questions should we ask in our mind? It doesn't work well to ask questions in a random way so make three axes and ask questions moving along the axes. First, try asking questions by moving the "time" in relation to the currently given subject. For example, you can ask, "What will happen to the subject 10 years later?" Second, try asking questions by moving the "space," say, what it is like in the U.S. or Saudi Arabia. The third axis refers to field. It is to try thinking by changing the sphere of a given subject. If a given subject is marketing, try thinking what this subject is like in the field of biology. Moving along the three axes like this can be regarded as asking questions traveling a three-dimensional world or moving along three navigations.

So what is a habit? As shown earlier, it is nothing but forming a neural circuit in the brain. Therefore, it only matters to create harmony among neurons. Persistent repetition simply produces teamwork and a neural circuit as well. Artistic performers or athletes practice over and over again because it is the process of building a neural circuit in the brain. If we ask the aforementioned three-dimensional questions repeatedly, it makes a circuit in the brain, which exactly means making a habit. As such, if we make the habit of asking our own questions in our mind, we are led to many new ideas and at last become a creative person.

1.5 Foresights of Future

Is it possible to foresee futures? Of course, we don't know well. But one thing for sure is that 20 years later things will be quite different from the present. Why? Compared to 20 years ago, things have become so much different today. Twenty years ago, there was neither Internet service at home nor cell phones. Just imagine a world without the Internet or a cell phone. Such things have been born in the past 20 years.

The future will change faster. We should not fix your eyes on things looking good today. It is important to look out on the change that will take place with a perspective of 20 years. However, that is easy to say but hard to accomplish being buried in reality. Indeed it seems not a good idea to plan the future for children after what is said to be good 'today'.

Today becomes the past in the future. We can prepare for what looks good today but 20 years from now, it will have been a thing of the past. Twenty years from now when these children are in their prime, I think the world will have been a global village without national borders. We should surmount the space of this country where we are now. Major disciplines in the limelight will change one after another.

There is no special secret in foreseeing the future. However, I see that continuing to observe the changes in technologies, environment, populations, and international relations will enable us to read the flow of the world. These are important factors of changing the world. To observe these changes, it is useful to view the world from the three dimensions of time, space, and field. We must look from three dimensions disengaged from the reality we are placed. As I said in the previous sections, we are unconsciously shut up in the 'wall' of reality. So we must 'break away from reality' by getting out of this wall.

Apart from reality, we can see the world from 'a bird's-eye view.' We can see much better than those who have their view blocked by a wall. Even what seems as a matter of course to those who are buried deep in reality looks different if seen from a distance. It is the same principle that applies at chess as a kibitzer reads the board better, which, if seen differently, will lead us to form 'questions.' These questions set us thinking freely.

If we are complimented at this time, we will come to ask questions even more often. Giving compliments is important for making habits. Thinking three-dimensionally can be called "a frame of thinking" that enables one to ask questions

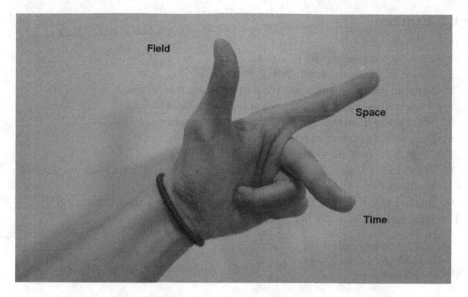

Fig. 1.6   Left hand law of creativity showing three dimensions

of one's own free will. It is also a three-dimensional "technique of solving problems". Besides this, we must not lose track of the given questions. Though it is not hinged on reality, we must keep asking the given question within these three dimensions and try to move our thoughts guided by the three-axis navigation to find a solution to a given problem. Creativity can be cultivated by extending our thoughts along the three axes: time, space, and field. Figure 1.6 shows "left hand law of creativity" showing three fingers pointing the three elements.

1.6   Summary

1. Our brain is stimulated when a question is given, and a new idea is likely to occur by questions.
2. We are fixed on reality and asking questions helps release us from reality.
3. A habit can be formed by creating a neural circuit in our brain.
4. A neural circuit is built through repetition which results a habit.
5. Giving compliment is important for reinforcement of the neural circuits.
6. Building a habit of asking questions to ourselves increases creativity.
7. Ask questions by extending thoughts along the three axes of time (T), space (S), and field (F).
8. Cultivate the habit of asking questions in three-dimensional TSF.
9. Taking the habit of asking three-dimensional questions can make a creative person.
10. By knowing the mechanism of creativity, we will not simply give up and say that being creative is natural.

1.7 Exercise

1. How do you imagine the future in 20 years?
2. Please imagine your life in Africa forest?
3. Suppose you are a musician. What can you learn from electronic field?
4. What is the relationship between compliments and habit?
5. What is the relationship between compliments and neural circuits?
6. How neural circuits are related to creativity?
7. What is the method to foresee futures?
8. Why TSF (Time, Space, Field) bird's eye view is important to develop creativity?
9. Please explain the left hand law of creativity to show s three dimension?
10. Please design your life in cc three dimensions.

Chapter 2
Freedom from Time

Abstract Humans stick to a current situation unless a new stimulus occurs. Without any external stimulus to brain, a person would focus on or stick to the moment, and the mind would remain in its current location. Question can stimulate the brain and break the current situation. In this chapter, asking questions on time is recommended, and which can be a navigator to extend our thoughts along the time axis. Three specific rules are presented regarding how to move on the axis or change the order of a sequence.

Keyword Creativity • Asking questions • Time • Time axis • Navigation • Extension of thought • Idea travel • Transpose • Tempo • Translation

As stated earlier, humans stick to a current situation unless a new stimulus occurs. Without any external stimulus, a person would focus on or stick to the moment, and the mind would remain in its current location. In terms of area as well, the person would stick to a certain area that interests him. Such a status quo hinders a person from thinking of something else, but the situation changes as an external stimulus affects him. A question, for example, may become an external stimulus to the brain, making the mind depart from the status quo and move to another area that the question leads into.

Here, the question can be spoken of as a three navigation system—a navigation system that leads a vehicle to movement. Likewise, the question described here functions to move our thought. A suggested question may be used when the mind has lost a train of thought and no new idea comes up.

How to ask a question on time will be illustrated here. This is a navigation system that moves the brain on a time axis.

K. H. Lee, *Three Dimensional Creativity*, KAIST Research Series,
DOI: 10.1007/978-94-017-8804-5_2, © Springer Science+Business Media Dordrecht 2014

2.1 Time Axis Traveling

A question on the time axis comes first. The time axis indicates the present moment and we are on the current point of time as illustrated in the Fig. 2.1. On this axis, you can see not only the current time but also that in 10 or 20 years. It would be good to move our thinking on this time axis. You may move your thought to 2024, 10 years forward, or 2034, 20 years forward. Turning back to 10 years ago, you may go back to the year 2004.

Performing this cannot be done automatically. When a problem occurs, you may move the time axis by asking a question to yourself. On a time axis, questioning can lead not only to movement but also change in the order of time.

As if driving a vehicle according to the guidance of a navigation system, the thinking is led by the time axis. Ask yourself how the given problem or situation may change as you go time traveling, so to speak, on the time axis.

This mind activity can be applied to almost every situation. Think about a paper cup. You want to make a disposable paper cup that is convenient, cheap, and eye-catching. What would you do?

As stated earlier, think as you move backward on the time axis. To free your mind, you need to break free from the present moment. Would your paper cup still be used 30 years from now? This "question" might hit you if you are not bound by time. Think of how the society might change in 10 or 20 years. A new idea may come up if you think of what kind of disposable cups would be used in such a changed society.

- A plastic cup that degrades?
- A cup that is foldable?
- A cup that is edible after use?
- A cup that is hygienically in good condition even after it was used?

There could be a lot of interesting ideas in addition to those above.
Let us say that we are thinking of a marketing strategy for vehicle sales.

- Would the life pattern of people in 20 years be the same with that now?
- What would the society in 20 years be like?
- How would habits of using a vehicle change as the society changes?
- Would the use of vehicles still be limited merely to transportation in time?
- How would the time that people spend in vehicles change?
- Would people drive with their hands on the steering wheel like today?
- What would people do in a car?
- How would marketing and promotion media change?
- Would people buy a vehicle in cash?
- Would cash be the major means to pay in the future?

Fig. 2.1 Time axis

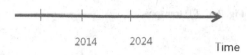

2014   2024

Time

2.2 T1: Transpose

Three specific rules are presented regarding how to move on the axis or change the order of a sequence. A question may be asked according to the suggested three rules.

The first is the Transpose rule called T1. We may think of a certain event according to time passage from the past to the future, or from the present to the past. In this rule, a question associated with order change may be asked when a problem arises. Time is changed on the time axis in other words.

A question may be asked in various perspectives as follows.

- Reverse—what if the flux of time is reversed?
- Periodic—what if the flux of time is periodical or irregular?
- Continuous—what if the flux of time is continuous or discontinuous?
- Pre-action—what if works are processed earlier than planned?
- Re-sequencing—what if the order is resequenced?
- Skip—what if a certain step in the order is omitted?

A further explanation is added below with some figures. Figure 2.2 shows a laundry machine. State-of-the-art laundry machines reset the program based on the extent of dustiness prior to cleaning. By using the corresponding buttons, the machines may be adjusted to reset such options as how long the cleaning will last, how much water will be used, how high or low the temperature will be, how many times the clothes will be rinsed, or when the spin-dry will be initiated. This is one example of re-sequencing of the laundry order in application of the Transpose rule. The question "What if the Transpose rule is applied to existing laundry machines" is asked here. When the order and length of washing steps are changed, different results occur and new products are developed in reflection of this finding.

Figure 2.3 shows ramyeon noodle. How would you cook ramyeon? Mostly people would boil water first. When the water is boiled enough, you would put noodle into it. When the noodle is cooked, you would add vegetable base, powder, and egg. The cooking ends then. Here, you may ask the question "Does ramyeon always have to be cooked in this order?" For example, you may put in powder first and then the noodle. Simply changing the order of cooking ramyeon may alter its flavor. If the resulting change to the flavor is positive, an added value is revealed. This is another example of applying the T1 or Transpose rule (Fig. 2.4).

In general, there is a certain order to cooking noodles: boil noodles in the water; throw out the water and transfer the noodles to another vessel; make some soup or sauce for the noodles; add the soup or sauce to the noodles once it is ready; the

Fig. 2.2 T1: transpose

Fig. 2.3 T1: transpose

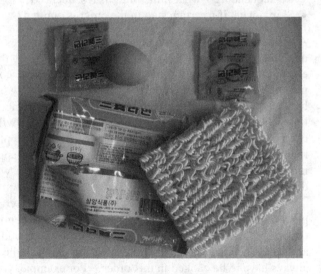

cooking is completed. Here, one question may arise. Why do boiling noodles and making soup have to be separate? Could the order of boiling noodles and making the sauce be changed in some way? You may change the order after asking if cooking should be done a certain way necessarily. Why is the water used to boil noodles always thrown out? What if noodles are boiled in the water just as in cooking ramyeon, and the water is used as the soup and as part of the finalized dish?

Fig. 2.4 T1: transpose

Fig. 2.5 T1: transpose

Another example is sushi. Sushi, soy sauce, and horseradish are shown in the Fig. 2.5. There could be different orders to eating sushi. For instance, you may add some horseradish in the soy sauce and then dip sushi in the soy sauce containing the horseradish. You may also dip sushi in the soy sauce without horseradish and put some horseradish on the sushi once you eat it. There will be some difference in taste between these two approaches. The order of adding horseradish and soy sauce makes the difference. This is another added value in application of the Transpose rule.

Figure 2.6 shows the dual spiral structure of DNA, a gene of living creatures. This is a structure that connects the four bases—A, T, G, and C, each of which is arranged in an order. The phenomenon of living creatures is decided by their order. Different orders result in different phenotypes. When DNA is inherited from the parents, this order might change. The crossover (change of the order) when the DNA is reproduced could be the cause. Further, a mutation might remove a certain part of DNA or add a new element to it, changing the order of DNA and causing a new phenotype.

At the bottom of the figure below, a genetically-altered GMO corn is shown. A large percentage of corns that we eat today are those that are genetically altered. These are of a certain species of corns that were artificially made for higher productivity and resistance to diseases. Mass production for food wealth is a means to solve the problem of food shortage. This is another example of applying the Transpose rule.

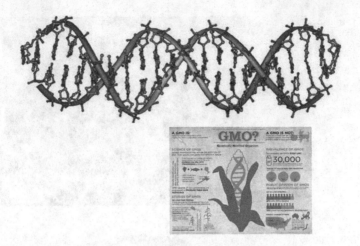

Fig. 2.6 T1: transpose

Fig. 2.7 T1: transpose (*Left* is ABS, and *right* is non-ABS brake)

Another example of applying this rule follows: The road might be slippery for drivers when it is snowing or raining. You need to be careful especially on a slippery road with curves. In such a case, you would step on the brake to reduce the speed. Stepping on the brake of common vehicles may cause the vehicle to totter and for it not to stop instantly. To prevent this from happening, a brake system called ABS has been developed. Common brakes keep the wheels stopped while ABS stops them discontinuously so that the vehicle does not totter but stops properly. This is another example of applying the Transpose rule that involves the question stated above. We may think that a brake is supposed to stop the wheels continuously. ABS applies the principle that holding discontinuously prevents the vehicle from losing its direction and places it on the right track (Fig. 2.7).

Fig. 2.8 T1: transpose

$$Y = -2x + 1 + x^2 \qquad (1)$$
$$= x^2 - 2x + 1 \qquad (2)$$
$$= (x - 1)^2 \qquad (3)$$

Fig. 2.9 T1: transpose

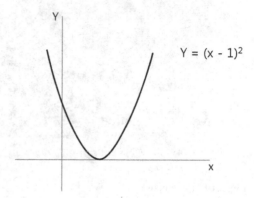

Figure 2.8 includes numerical expressions. Expression 1 is $-2x + 1 + x^2$. It is difficult to reason the relation between the value x and value y. Let us go to Expression 2. The order of Expression 1 has been changed to $y = x^2 - 2x + 1$. This expression might be changed to Expression 3 as well; that is, $(x - 1)^2$. This will help you to deduce a two-dimensional curved line. Such is also an example of applying the Transpose rule to numerical expressions. A new idea was drawn out by changing the order of terms. The relation between x and y is visually represented by applying the Transpose rule (Fig. 2.9).

2.3 T2: Tempo

This section presents the second rule of time, which is T2 or tempo. Tempo indicates speed. A new idea may be drawn out by questioning the speed change on the axis of time. The questions below might be asked.

- What would happen if the speed increases?
- What would happen if the speed decreases?
- What would happen if the speed is adjusted in a uniform rate?
- What would happen if the speed gradually increases?
- What would happen if the speed gradually decreases?
- What would happen if the speed changes irregularly?

Figure 2.10 illustrates bicycle riding. When the speed of the bicycle that you are riding is close to 0, it is likely to fall. A bicycle needs to maintain a certain level

Fig. 2.10   T2: tempo

Fig. 2.11   T2: tempo

of speed for it not to fall down. This bicycle has been invented in reflection of this principle. The T2 or Tempo rule was applied with the questions above in inventing the bicycle.

Figure 2.11 show an example of gaining a newly added value by increasing speed. The figure on the left side shows someone figure skating in a beautiful posture, which cannot be done unless a certain level of speed is maintained. A variety of beautiful postures can be presented only when the skater maintains a certain level of speed. Roller boarding is another example. A roller board needs to

Fig. 2.12   T2: tempo

Fig. 2.13   T2: tempo

maintain a certain level of speed to exhibit routines as in figure skating, both being examples of adjusting speed.

Figure 2.12 shows a top spinning. The top would stop turning if you do not spin it. It would continue spinning if it maintains a certain measure of speed, which is another example of gaining added values by changing the tempo.

Figure 2.13 is a Drop Zone in the Children's Park. Think of this: Who would ride this machine if the Drop Zone comes down slowly? This is another added value produced by applying a sudden fall or free fall. This is another example of applying the Tempo rule.

Figure 2.14 shows an airplane landing on the aircraft carrier. The most difficult moment for an airplane taking off is landing. As you may know, the runway where an airplane lands on is relatively short; thus, the technique to land on this short runway is of great importance. If the airplane does not stop properly on it, the aircraft will fall into the sea.

To prevent such an accident from happening, a hook comes out of the back of the airplane when it lands as shown in the figure. This hook is linked to one of

Fig. 2.14   T2: tempo

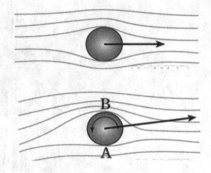

Fig. 2.15   T2: tempo

the four wires on the runway of the carrier for safe landing. When the hook is not linked to the first one properly, then hopefully it will to the second, the third one, or the fourth one. What if the hook is not linked even to the fourth one? The airplane would fall into the sea. Is there any method to prevent the worst scenario from happening?

In general, the pilot would speed down the airplane when lending. However, right before landing on the carrier, the airplane would maintain the highest speed. The airplane is expected to be linked to the wire on the runway so that it speeds down and lands safely. If it is not linked even to the fourth wire, the airplane can take off since it has maintained enough speed before trying to land.

This is an example of doing away with the fixed idea that an airplane is supposed to speed down when landing as it could speed up to the highest level to be able to land on a short runway safely.

Figure 2.15 shows a curveball that a pitcher throws. The ball that the pitcher throws changes direction as it moves forward. Such magic depends on the speed and spinning of the ball. The figure above shows a ball that moves ahead without spinning. The ball goes straight with no change in direction no matter how fast it is since it is not spinning. The ball in the figure below, in contrast, is spinning. If you look closely at the surface, the air speed at Part A is slow while that at Part B is fast. Pressure goes down when an object moves fast. This is called Bernoulli's Principle. The ball changes its direction toward the part where it receives less pressure and which makes the ball curve. This is the magic of speed.

2.4 T3: Translation

The last and third rule of time is T3 or translation. A lot of available data is on a time axis. Such data is in direct relation to the passage of time. The Translation rule is an attempt to interpret it in exceptional situations. To do this, the data set is represented in graph. Such questions below might be asked about whether there are any new exceptional situations on the time axis:

- Is there any exceptional data?
- What form of graph does the exceptional data make?
- What form of graph does the exceptional data turn to?
- Does the exceptional data increase or decrease in quantity?
- Is there a certain pattern for such an exceptional data?
- What should be done to turn the data pattern to your desired shape?

As shown in the Fig. 2.16, the data is represented as a graph in the time axis. The data rapidly increases in quantity and forms a graph. In general, a sudden, abnormal phenomenon to the data is likely to be neglected as a mere mistake in a certain experiment or measurement. However, questions need to be asked on such an occasion. An unexpected fact might be found depending on how you interpret such exceptional, suddenly emerging data. This exceptional data might reveal a secret that was not expected from the data.

In the graph (Fig. 2.17) where the values of data gradually increase, whether such movement involves a certain trend of gradual increase or decrease needs to be addressed. In addition, whether such a phenomenon involves any new, unexpected factors needs to be examined in application of T3 or the Translation rule.

Alexander Fleming's discovery of the penicillin is one of the examples of such a translation. When Fleming went on his vacation, he left the bacillus plates outside. After coming back from his vacation, he found that the experiment failed due to the bacillus reproduction, but he happened to observe that a certain part of bacilli died as he was about to throw them away. He found blue molds around that part. If he neglected this exceptional data, it might have taken a lot longer before the penicillin was released. Fleming did not simply take for granted the dead bacilli where there were blue molds. He figured out that blue molds functioned to kill bacilli after investigating the cause for a considerable period of time. Such efforts resulted in the development of penicillin which made a remarkable contribution to mankind (Fig. 2.18).

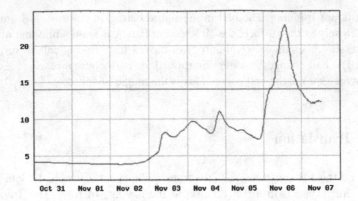

Fig. 2.16   T3: translation

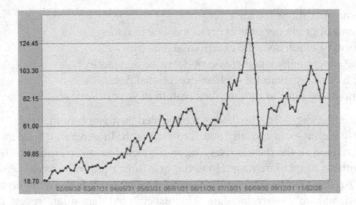

Fig. 2.17   T3: translation

Fig. 2.18   T3: translation

2.5 Designing the Future

Everyone dreams of success in the future and tries to realize the dream that they have. When do they expect their dreams to come true? Tomorrow? The next month? In 5 years?

Most people who dream of success would not expect it to happen right away, and it could not. The periods that they expect to take might vary, but it certainly is a prolonged period of time, which could be 5 years or 10 years. Young students in particular may expect that to happen when they have fully grown. A 15-year-old student may imagine success to happen when he becomes 35 years old, which is about 20 years from now.

How would you prepare if you expect success in 20 years? Certainly, you would imagine the 2030s, when you will be realizing your dream. This frees you from the bounds of reality. In other words, asking yourself the following questions will free you from the limits of reality:

What would the world be like in 20 years?
What kind of mobile phones would be used in 20 years?
What kind of vehicles would be used in 20 years?
What would schools be like in 20 years?
What kind of TV would we watch in 20 years?

The questions above prompt us to travel on the time axis. When we leave the world to 10 or 20 years from now, we can free ourselves from "today" where we are bound. Being freed from the reality of today, you can give full play to your imagination with much more fresh thoughts. People thinking of the future will be strengthened to overcome the difficulties of reality that they are facing. It will become easier to overcome current trials if they look forward to good results.

There is a book entitled *Don't Eat the Marshmallow... Yet!* written by Joachim de Posada and Ellen Singer. The basic idea is that if you save the marshmallow that you want to eat today, a bigger reward will follow tomorrow. This encourages you to patiently resist the temptations of reality, wait for rewards, and prepare for the future. The same applies to the point addressed here.

If today were the only day that you have in your life, you better eat up the marshmallow right away. But if you look further into the future, you may find that you can eat a lot more marshmallows if you practice moderation now. Controlling your desire to eat the marshmallow right in front of your eyes is in line with developing your skills for the future. If you make it a habit to look to the future in 10 years, suffering at the moment may seem less difficult to overcome.

This applies to when you argue and get angry at someone. When you are exasperated, you may want to vent your wrath in some way, but you can compose yourself if you think of the future after one year since what just happened would be nothing important by that time. If you move yourself to another point on the time axis, you can calm down after the outburst of emotions. People may recover their composure by moving themselves on the time axis.

2.6 The Founding of Nexon

Nexon is an Internet gaming company that released such popular games as *Kingdom of the Wind, Maple Story, Kart Rider, Dungeon Fighter, and Mabinogi*. This group is known as one of the the world's top three Internet gaming businesses next to Nintendo and Activision Blizzard. Nexon has more than 3,000 employees, and its annual sales reaches 1 billion dollars. About 0.3 billion individuals in 140 countries around the world are using Nexon's games.

China accounts for 37.5 % of the entire sales of Nexon, with Korea 32.7 %, Japan 14.3 %, and the U.S. 8 %. According to Forbes, a magazine in the U.S., the private properties of the founder, Jeongju Kim, reached USD 2 billion early in 2011, the 595th highest in the world. Late in 2011, as Nexon's stocks were listed in the Japan stock market, its properties increased to USD 6 billion.

Jeongju Kim, the founder and chairman of Nexon, was an eccentric student with yellow hair who would daydream in my research laboratory at KAIST about 16 years ago. He would say that he was developing a program through which people far from each other could play a computer game. I told him:

> Computers need to be linked through a network for graphic images to be delivered between stations real-time. Networks need to have enough speed for real-time games too. Your idea may not be realistic.

Back then, the Internet was not widely accessed, and there was no high-speed communication network through which images could be transmitted real-time. Only telephone lines were used for PC-based communication. He replied:

> Professor, in about ten years, every computer will be linked to a super high-speed communication network. I will make computer games that can be used in that time.

A brick-breaking game called Tetris was popular back then. It needed no network since it was a person-to-computer game. The fact that a human can play a game through a computer was innovative. Thus, his idea to make network-based computer games through which people far from one another could play seemed to be quite imaginary and unrealistic. Many were thinking of making Tetris II, an upgrade of the original version of Tetris, which was popular before.

Not "the present" but "the future" in 10 or 20 years was in his mind. With a map for the future 10 years ahead, he was shaping what was supposed to be the future of gaming. People around him, stuck in reality, could not understand his idea. They considered him an odd student. He attended classes during the day and developed games at night. Finally, he released his first product *Kingdom of the Wind*.

As the 21st century began, ultra-speed communication networks started to be established. Many started thinking of Internet games. Since Jeongju Kim's team started earlier, it is no wonder he burst ahead of others in this area. This is a result of moving oneself 10 years ahead on the time axis.

2.7 Summary

1. Whether we notice or not, we stick to the current point of time. When a question is asked, we break our attachment to reality.
2. When given a question on time, we move our line of thought to the point of time that the question addresses.
3. If we move along on the time axis, our thoughts can be freed from reality.
4. Rule T1: Transpose

- Reverse—what if the flux of time is reversed?
- Periodic—what if the flux of time is periodical or irregular?
- Continuous—what if the flux of time is continuous or discontinuous?
- Pre-action—what if works are processed earlier than planned?
- Re-sequencing—what if the order is re-sequenced?
- Skip—what if a certain step in the order is omitted?

5. Rule T2: Tempo

- What would happen if the speed increases?
- What would happen if the speed decreases?
- What would happen if the speed is adjusted in a uniform rate?
- What would happen if the speed gradually increases?
- What would happen if the speed gradually decreases?
- What would happen if the speed changes irregularly?

6. Rule T3: Translation

- Is there any exceptional data?
- What form of graph does the exceptional data make?
- What form of graph does the exceptional data turn to?
- Does the exceptional data increase or decrease in quantity?
- Is there a certain pattern for such an exceptional data?
- What should be done to turn the data pattern to our desired shape?

2.8 Exercise

1. What would result from asking a question in application of T1 rule to the menu order of my mobile phone?
2. Which rules (T1, T2, T3) could be applied to the process of factorization?
3. Which rules (T1, T2, T3) could be applied to the motions of riding a bicycle on a curved road?
4. Which rules (T1, T2, T3) could be applied to the motions of dismounting a bicycle?

5. Think of the process of making a hamburger. What change would result from applying the three rules of time (T1, T2, T3) to this?
6. Think of drinking water in an artificial satellite in zero gravity. What change would result from applying the three rules of time (T1, T2, T3) to this?
7. Analyze the process of tornado formation. What could be done to prevent such a tornado from being formed?
8. Think of the motions to catch a fly. What could be done to catch it fast before it flies away?
9. Evaluate mathematical the expression $y = -4x + 4 + x^2$ by applying the rules.
10. If we do not put case of a pen for white board in class room, it's ink is often dried. What kind of questions can we give to protect such dry out?

Chapter 3
Freedom from Space

Abstract Our thinking sticks to reality unless there is any external stimulation. Thinking is fixed on a current point of time, a current position, and an current area. A fresh idea is likely to come up when our thinking is freed from current situation. In this chapter, space navigator is introduced to extend our thought along the space axis. The spatial navigation system lets an individual that sticks to the current location or shape move to another place or shape. Three types of rules are provided to make question on space axis.

Keywords Creativity • Space • Thought navigator • Shape • Site • Size • Space travel • Location change • Position change

In the previous section, we have considered the navigation system through which we can move on a time axis. This section introduces the second navigation system, through which we can make a spatial movement. The spatial navigation system lets an individual that sticks to the current location or shape move to another place or shape. When given questions such as what would happen if the location shifts to another point or what would happen if a shape transforms, you may be motivated to seek changes in terms of space.

3.1 Space Travel

Let's call to mind the disposable cup considered earlier. We are trying to make a new disposable cup, but a new idea will not turn up. Ask questions then with the location on the spatial axis changed:

- Would the paper cup be used in another country just as it is?
- What type of disposable cups would be used in Alaska?
- What kind of beverages would people in Alaska drink?

K. H. Lee, *Three Dimensional Creativity*, KAIST Research Series,
DOI: 10.1007/978-94-017-8804-5_3, © Springer Science+Business Media Dordrecht 2014

Fig. 3.1 Space axis

Space

USA

China

EU

Korea

- What kind of cups would be appropriate in Saudi Arabia?
- What kind of cups would be appropriate in jungle areas in Africa?
- What kind of cups would be appropriate in ships?
- What kind of cups would be convenient in airplanes?

Thinking of the answers to these questions, we may come up with a new form for paper cups. While imagining different locations and environments, a new idea may appear (Fig. 3.1). Now the three rules—S1, S2, S3—are introduced below, which will help us to generate questions.

3.2 S1: Shape

The S1 rule involves questions on different shapes of a certain object. Specifically, changes in terms of formation in a certain space are addressed.

- If a certain part of an object is lineal, what if it changes as curved?
- If a certain object is symmetric, what if it becomes nonsymmetric?
- If a certain object consists of dotted lines, what if the lines turn solid?
- If a certain object is flexible, what if it is nonflexible?
- If a certain object is flat, what happens if it becomes nonflat?

To illustrate, figures are used below. Figure 3.2 shows a kitchen knife used in common households. A knife usually has a sharp edge. When this edge becomes blunt, a whetstone may be used to resharpen it. On the right side of the figure, the edge of another knife is ridged just like the teeth of a saw. Is it not weird? It is not ordinary that a knife has saw teeth. The knife shown in this figure has saw-like teeth so that there is no need to resharpen the edge. It is difficult to sharpen knife edges by

Fig. 3.2 S1: shape

Fig. 3.3 S1: shape

means of a whetstone. This knife was designed to have teeth instead of a common edge so that it can be moved back and forth like a saw. The knife would work better when it moves back and forth. This alteration in shape enhances the existing knives that need to be resharpened by means of a whetstone.

Look at the figure below. In general, people expect objects to be symmetrical. We tend to make things symmetrical without even noticing it. This knife has saw teeth at one side and none at the other side. This knife is nonsymmetric, which departs from the unconscious intention to make things symmetrical. This goes beyond the scope of symmetry, which can result in extra added values.

Figure 3.3 shows kitchen scissors commonly used in families. Scissors are convenient when used in a kitchen, but there is one disadvantage to it. Over time, foreign substances may be stuck at the back tooth of the scissors or the screw connecting the two edges. Thus the scissors may eventually resist to move properly. This model was designed in a way that the pivot area can be detached to remove the foreign substances. The form of the part connecting the two edges has been changed so they could be detached. The figure shows the edges separated.

Fig. 3.4 S1: shape

Fig. 3.5 S1: shape

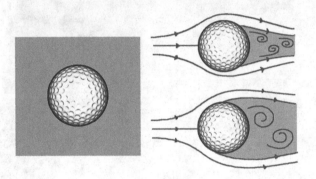

The teeth of the scissors have been changed as well to a round shape where objects are fit for cutting so that the object does not slip. This is an example of gaining added values by changing a shape. Two shape rules are applied to the scissors. First, the rivet connecting the two blades is designed to be unfastened. Second, the teeth of the scissors are redesigned so that an object becomes better fit for cutting.

Let us consider another way to apply the S1 rule. Figure 3.4 shows the inlet of a vacuum commonly used in a house. The left side shows the common model while the right side shows one that can be used to clean corners or books. This is an example of designing a new function by changing a shape.

Gaining added values by changing a shape is applied to golf balls (Fig. 3.5). The ball needs to go as far as possible once it is hit. The surface of the ball is designed to be smooth for this reason. However, the surface of a golf ball is not smooth but rough. This texture reduces the vortex from behind when the ball is moving forward and causes the ball to go farther. Changing the ball's shape has created this function.

Fig. 3.6 S1: shape

Fig. 3.7 S1: shape

The next example is a ladle (Fig. 3.6). A ladle is used to scoop the soup into a bowl. Sometimes solid ingredients must be scooped with the soup. For this purpose, some ladles have sawlike teeth so that soup and solid ingredients such as noodles can be scooped. In this case as well, changing the common shape of the ladle creates an added purpose.

There are still many other interesting examples of applying the S1 rule. Some scissors are designed to be used in eating crabs (Fig. 3.7). In general, the length of the edges of scissors is the same at both sides, but in this case, one side is longer than the other, and the longer side is sharp so that you can pick and eat the inner meat in the crab legs. This is an example of changing the shape of scissors specially for crab eating.

Figure 3.8 shows a urinal unit in a men's room. Men are likely to splash while urinating. If there is a target or picture at the drainage of the urinal, people may try to aim at the target just as they would extinguish fire unconsciously in a sighting shot training.

Fig. 3.8 S1: shape

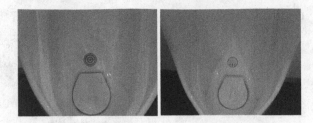

Fig. 3.9 S1: shape

Figure 3.9 shows vehicles from the upside. In general, people tend to pursue vertical symmetry, and this is true for most vehicles. The vehicles in the figure are not of vertical symmetry. There is one door at the left side while there are two at the right side. The design is based on the fact that the door at the left side is hardly used. This asymmetric design contributes to saving costs and lending uniqueness to a shape.

Figure 3.10 shows an airplane taking off an aircraft carrier. Gaining added values by changing a shape is seen in this sample. A jet moves forward through the momentum of the engine, which emits air behind. Since the runway of an aircraft carrier is relatively short, more pulling force is required. The force to emit air at the back of the jet is required more than in other cases. When the airplane takes off from the runway, as seen in the figure, a type of board is lifted at the back of the carrier's runway so that the air from the jet engine hits the board and thus the airplane goes faster. This way, jets can take off even on a short runway.

Figure 3.11 shows tires. Various patterns are designed on these tires. The pattern at the left side is called rib. This rib type pattern seems to be in the direction of the vehicles as shown in the figure. This contributes to a vehicle's forward movement without slipping to sides and generating excessive heat. In contrast, the figure at the right side shows lug type tires. Unlike the rib type, the lug pattern on the tire surface is horizontal and right-angled to the vehicle. This is advantageous considering the force of moving forward and braking. This pattern is commonly used for trucks because of its braking force and heat emission. The pattern of the

Fig. 3.10 S1: shape

Fig. 3.11 S1: shape

third tire is hybrid and asymmetric. Some parts of the pattern are in direction of the vehicle while some are right-angled. The horizontal pattern is asymmetric for stronger force when the tire contacts the road surface. In addition, it prevents the tire from being worn out, enhances the braking performance, and contributes to the grounding at the outer side of the tire when the car is cornering. These are some of the examples of changing the shape of a tire.

Figure 3.12 shows three different canoes. The first one shows a common type of canoe. This canoe is likely to fall down to the side since the center of gravity is high. Natives, who have used canoes since long ago, tried to secure stability by adding a wing to the canoe. In the figures below, the canoe has a wing at the side. In order to make a canoe with a wing fall down, we have to pull up the center of gravity. Pulling up the center of gravity is difficult because of the wing. As a result, the canoe becomes stable.

Fig. 3.12 S1: shape

3.3 S2: Site

This section explains the second rule applied on a spatial axis; that is, S2: site or location. The questions below are about changing location in space.

• What if an object is relocated to the upper side?
• What if it is relocated to the lower side?
• What if the center of gravity is moved to the lower side?
• What if the center of gravity is changed to a higher or lower level?
• Will the object be located in a local area?
• Will the location be globalized?
• What if a certain location (part) is emphasized?

Figure 3.13 shows an example. This figure shows a roly poly flowerpot. When the flowerpot is watered and wet, the flowerpot is straight while it may be inclined when dry. This indicates when it requires to be watered. The idea is to change the center of gravity according to the humidity that can be applied to a system that gradually reduces the water.

Figure 3.14 shows a child playing with a toy. This is a type of scale whose inclination is adjusted depending on the location of the pivot or the center. It is balanced by relocating the pivot.

Figure 3.15 shows the TV on the wall in the author's office. If you look closely at the screen, the person is upside down because the TV set is upside down. I have been watching TV upside down for the last 6 years. This has helped me get out of fixed ideas and think in different perspectives. I watch TV about 10 min a day with the TV set upside down. This is a change that the author made after questioning himself, "What if the TV set is turned upside down?"

Figure 3.16 shows how to stand up with little load on the waist while holding heavy things. You have to keep yourself down when picking up the object. If you keep the body's center of gravity down, you can prevent your back from being hurt. This is an idea of changing the site or the center of gravity.

Figure 3.17 here shows a washing machine on the wall. In general, washing machines occupy a lot of space. Since the living space is small for many

Fig. 3.13 S2: site

Fig. 3.14 S2: site

Fig. 3.15 S2: site

people, this was designed to reduce the space that a washing machine occupies. The location of a washing machine has been changed in the case of this washing machine model installed on the wall. The wall space is utilized as the washing machine is relocated from the ground to the wall. This is an example of changing the location.

Fig. 3.16 S2: site

Fig. 3.17 S2: site

This is an example of numerical formulas (Fig. 3.18). Formula 1 has the root of 20 for the denominator. It is somewhat difficult to guess the value of y so the location of the root is changed. As in the case of Formula 2, $\sqrt{20}$ is changed to $\sqrt{5}$. As in Formula 3, $\sqrt{5}$ is multiplied by the numerator and denominator at the same time to increase the root to the numerator. With Formulas 4 and 5, you may readily guess that the value of y of the numerical formula is $\sqrt{5}$. As for Formula 1, it is difficult to guess which is bigger. Relocating the root makes it easy to know that the value is $\sqrt{5}$. This is commonly used in mathematics.

Fig. 3.18 S2: site

$$y = \frac{10}{\sqrt{20}} \qquad (1)$$

$$y = \frac{10}{2\sqrt{5}} \qquad (2)$$

$$y = \frac{5}{\sqrt{5}} \qquad (3)$$

$$y = \frac{5\sqrt{5}}{5} \qquad (4)$$

$$y = \sqrt{5} \qquad (5)$$

Fig. 3.19 S2: site

Figure 3.19 shows a hydroelectric power plant. A hydroelectric plant generates energy when water falls down. Relocating water converts potential energy into electric energy. This is an idea in application of Site rule.

Let us consider the example of oil tankers (Fig. 3.20). An oil tanker runs on oil but sometimes tankers may run with no oil. Afterwards, the center of gravity becomes higher and thus the tanker may overthrow. For this reason, an oil tanker needs to be filled with water to move its center of gravity down. This is an application of the idea of changing the center of gravity.

Fig. 3.20 S2: site

Fig. 3.21 S2: site

In Fig. 3.21, there is a screen of ATM machine of bank. When we input our password number to touch screen monitor, there is a possibility that other people see our finger movement for the secret number. In the figure, we can see the positions of numeric numbers are not same with the conventional one. Each time the monitor gives number screen, the positions are different, and the other people cannot image our secret numbers by our finger movement.

3.4 S3: Size

This section introduces the last rule of space. On the spatial axis stated above, S1 changes the shape while S2 changes the site. S3 is a navigation that changes size to produce different ideas.

- What if the object is reduced?
- What if it is enlarged?

Fig. 3.22 S3: size

Fig. 3.23 S3: size

- What if it is thicker?
- What if it is thinner?
- What if the thickness is not uniform?
- What if the weight is reduced?
- What if the weight changes?

This is a semiconductor chip (Fig. 3.22), which is utilized in almost every device used today. Some gigabyte electronic component is inserted into this small chip, which is an example of downsizing. Almost all semiconductor researchers and producers around the world are trying to downsize electronic components and put them into a more compact circuit.

Figure 3.23 shows cell phones which are getting slim and light recently while the screen is getting larger. This is an example of gaining added values by adjusting the size.

The S3 rule is also applied to TV sets (Fig. 3.24). People want their TVs to be lighter and thinner as the screen gets larger. To design TV sets this way, the external frame needs to be thinner. The competition in the TV industry as well is fierce with regard to changing the size.

Fig. 3.24 S3: size

Fig. 3.25 S3: size

Figure 3.25 shows a very small refrigerator. This is a refrigerator model for cosmetics, which goes beyond the fixed idea that refrigerators need to be big. The size is small because this model is specially designed for cosmetics. It is so small that it can be placed even on a dressing table. The wall-mountable washing machine considered earlier with regard to the S2 rule is an example of downsizing. Changing a size and weight in application of S2 and in combination with changing the location led to creating a wall-mountable model.

3.5 A Lesson of Cyworld

Cyworld is a company initiated in Korea by Yongjun Hyeong, a graduate from the Department of Business Management, KAIST, with his colleagues in 1999. It was originally designed to embody the trust-based information-sharing system that he was researching in a lab. Donghyeong Lee and others joined and expanded the business since 2001. In 2003, it was integrated into SK Communications. Its number of members in Korea once reached 10 million which corresponded to one quarter of the Korean population. This pioneer of social networking service has recently been overcome by Facebook.

It is worth thinking why Cyworld declined. Compared to Cyworld, Facebook was created very late. Zuckerberg, a student at Harvard, designed it in 2004 just for fun. Now the number of Facebook members is over 1 billion around the world. What made it more distinguished?

There could be many reasons, but the biggest one was Cyworld's regional limitation or its Korean culture roots. Cyworld attempted to globalize itself after its domestic success. Unfortunately, though, the service did not attract foreigners much since it was too Korean. Service products are completely different from hardware products such as TVs and ships, which involve merely follow-up repair services after sales. In contrast, service products need continuous maintenance. Because of inevitable constant interaction, cultural characteristics are involved.

As we all know, Japanese, Europeans, Chinese, and Americans have different favorite colors. They are likely to have different favorite colors and designs in Internet home pages as well. Accordingly, the home page menu designs, services, and interfaces need to be different.

You may find that the home pages of global companies are attractive to anyone. This is because they design things in a globalized perspective. In application of S2, they go beyond the limits of location in designing things.

In utilizing smart phones, such factors as design of buttons on the screen, number typing methods, order of finding numbers, and fingering methods are an interface or a language between humans and machines. A language is a culture. Favorite interfaces may differ depending on the culture. Steve Jobs created outstanding interfaces that almost all in the world would accept without reservation.

What about Facebook? The service is accepted throughout the world. Its home page was designed to attract various types of viewers. No wonder it is used worldwide.

3.6 Globalization of Nexon

Internet games as well endeavored to become globalized. Nexon, an Internet game business discussed earlier, began its service in Korea but relocated its head office to Japan, a gaming center, to develop its services later. It aimed to reflect globalized senses right from the step of planning. Nexon advanced well into overseas markets probably for this reason. 70 % of its sales is from overseas markets. This is an example of successfully going beyond the limits of space as a result of S2 or Site questioning.

In addition, Nexon pursued localization strategies. It selected and promoted certain games depending on each country's cultures and consumer tastes. For example, users in Japan count cooperation, rather than competition, as important. This is reflected in RPGs (Role Playing Game) such as Maple Story, Tales Weaver, and Dungeon Fighter which are specially promoted in this country. RPG games do not encourage competition with others. Each gamer has his own role and enjoys performing it instead.

In the U.S., the company focuses on familiarizing the consumers with "partial charging." The game itself is provided for free, but the user may choose buying game items for more fun, to which the concept of partial charging is adopted. For instance, items to decorate the game characters are available. The partial charging system has been established especially for simple casual games such as Cart Rider and Maple Story.

In Europe, games that require advanced specifications such as Mabinogi Heros are promoted rather than simple casual games. Users in Europe prefer serious games with magnificent graphics and a variety of missions to simple games.

3.7 Globalized Thinking

Things need to be designed in a way that can attract people around the world. The importance of cultural elements is not recognized much when items such as ship, semiconductor, and TV are sold. As far as smart products with a user interface are concerned, however, cultural differences count. We need to adapt ourselves to global standards.

K-Pop music (Korean pop music) popular in Europe and the U.S. is a major example of successful globalization strategies. Rather than insisting local styles of writing, dance, and stage equipment, favorites of people in other parts of the world are considered and reflected in the planning. The popular music of Psy, for example, makes people in the world laugh and have fun all together.

There are many successful cases of broadening horizons in developing products. One of them is exemplified by mobile phones produced by LG. Its models are popular in Islamic countries because of the special functions available for the culture.

Muslims pray five times a day in the direction of Mecca. Wherever they are in the world, they find the direction of Mecca, kneel down, and pray toward it. With this fact in mind, LG mobile phones have designed and included functions to find the point of Mecca so that Muslims can pray toward it.

A new environment is created once you leave the current region and proceed to another. Ask yourself questions regarding another region. Question yourself while moving along the spatial axis. You may ask, "How would the problem given to me be handled in China?" "How would Americans use a product?" "How would Muslims respond to a feature?"

You can free yourself from your current spatial limit if you think this way. You can free your mind and come up with original ideas.

Amy Chua at Yale University studied the rise and fall of world powers such as Rome, Mongolia, Spain, America, and Great Britain throughout human history and stated her findings in *Day of Empire*. She concluded that tolerance counted in each ascension and descent. When foreign things and human resources are tolerated and accepted, a nation could be powerful and prosperous. In contrast, a powerful country might end up falling if it emphasizes purity and exclusivity.

Whether it is a nation, a business, or an individual, a globalized way of thinking is of great importance to introduce new elements and stimulate an entity, which may lead to creative ideas. This is the same with designing one's future. It is necessary to look out into the world rather than stick to a current position if you want to grow as part of the global community.

3.8 Summary

1. Free yourself in a space.
2. Liberate yourself from your current location and position.
3. There are three rules involving the space axis:

(a) S1: shape

- What if something lineal currently turns to be curved?
- What if something symmetric currently turns to be nonsymmetric?
- What if a dotted line turns to be a solid one?
- What if something flexible turns to be nonflexible?
- What if something flat turns to be nonflat?

(b) S2: Site

- What if something is relocated to the upper side?
- What if it is relocated to the lower side?
- What if the center of gravity is moved to the lower side?
- What if the center of gravity is changed to a higher or lower level?
- Will the object be located in a local area?
- Will the location be globalized?
- What if a certain location (part) is emphasized?

(c) S3: Size

- What if something is reduced?
- What if it is enlarged?
- What if it is thicker?
- What if it is thinner?
- What if the thickness is not uniform?
- What if the weight is reduced?
- What if the weight changes?

3.9 Exercise

1. Apply the three rules (S1, S2, S3) of space to the bicycle that you are using.
2. You would like to develop a vehicle that you can park in a smaller space. Apply the three rules of space (S1, S2, S3).
3. Vegetable corner in a supermarket is often open and vertically displayed while a refrigerator is operated. What kind of question can we give to save energy in this situation?
4. When displaying products in a supermarket, you would like to put a lot of items in a limited space in a way that customers can conveniently browse. What questions may you apply in this case?

5. Apply S1 to the ball point pen that you are using and design a new model.
6. Water is spilled out if the water bottle is fallen. How can we prevent such spill even though the bottle is fallen?
7. Apply S1 and S2 to the white board in the classroom and design a new shape.
8. Evaluate the expression: $y = 1/\sqrt{8}$ by applying the rules.
9. Wine glass is easily fallen down. How can we prevent such falling down?
10. If an object falls down from a 10-meter high space, it gains momentum. Which rules (S1, S2, S3) are related to this phenomenon?

Chapter 4
Freedom from Field

Abstract Modern society continues to advance and grow ever complicated, and we seek specialized knowledge and focusing on a certain area may narrow your width of thinking. That is, we may stick to certain area or reality. Asking questions on field is important to make our brain free from the reality. In this chapter, a field navigator is proposed to extend our thoughts along the field axis. Three practical rules for making questions are also given: function, fertileness, and fusion.

Keywords Creativity • Field • Thought • Function • Fertileness • Fusion• Navigator • Reality • Field travel

As modern society continues to advance and grow ever complicated, it becomes difficult to acquire considerable knowledge in many fields. When trying to cover a wide range of areas, you are likely to lack depth in understanding them. For this reason, people seek specialized knowledge. The problem, however, is you may turn negligent of other things while you are absorbed in a certain matter. Focusing on a certain area may narrow your width of thinking and hinder you from viewing issues in a diverse manner.

Perspective is applied to our thinking. An object that is nearby looks big while one that is far looks small. An object that is nearby looks important while one that is far looks less important. When thinking only from one's position, it becomes difficult to maintain balance in general. Effort needs to be invested in the examination of peripheral areas. We may need to direct our eyes to another field intentionally. When you are given a certain problem, new ideas will only form if you apply them in other fields. Why not leave a thinking route to navigate "fields"? Let's move on field axis as shown in Fig. 4.1.

K. H. Lee, *Three Dimensional Creativity*, KAIST Research Series,
DOI: 10.1007/978-94-017-8804-5_4, © Springer Science+Business Media Dordrecht 2014

Fig. 4.1 Field axis

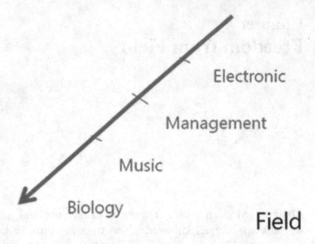

4.1 Field Travel

Here, a field may indicate a major or a business area. It indicates a sector that you are currently interested in. A field can be mathematics, physics, business management, arts, performance, credit card, schooling, exercise, smart phone, and so forth.

Steve Jobs has changed our life a lot. Was he a scientist that revealed secrets of nature like Newton or Einstein? Was he an inventor like Edison who created things that had not existed before? Jobs combined and converged things in various fields to create a new thing, which significantly changed the lives of people around the world.

Let us look back at the matter of designing the disposable cup stated earlier. Thinking of a new cup, we could conjure a different type of container since disposable cups made of paper are commonly used. Without even noticing it we limit ourselves into thinking only about this certain material, paper. But we need to think of the diverse possibilities of disposable cups.

- How about a cup with electronic features?
- A cup that indicates temperature?
- A cup whose color changes depending on the temperature?
- A cup that is edible after being used?
- Could the paper be made of materials other than natural pulp?

Three specific rules are applied here. The first rule, F1, is about function. The second, F2, is about material or fertileness. Fertileness indicates the wealth of potential materials, implying that it is recommendable to use a variety of materials rather than be tied down to just one. The third, F3, is a rule that puts special emphasis on fusion.

Fig. 4.2 F1: function

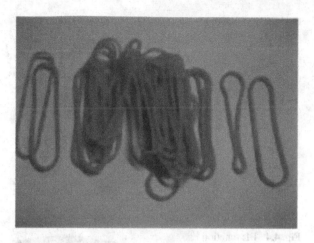

4.2 F1: Function

F1 addresses the functions of the given object. It seeks to adjust an existing function or replace it with another. Below are sample questions:

- What if the current function is changed?
- What if the given function is utilized for something else?
- What if something else is used to perform the function?
- What if the form of functions is changed to look different?
- What if one certain function performs various tasks?
- Would it be possible to adopt the functions of objects used in other fields?

Now let us consider some examples while referring to the figures below:

Figure 4.2 shows rubber bands. Their original usage is to tie something up. However, they may be used for many other purposes. A rubber band may be used for slinging, for producing interesting sounds, for preventing something from slipping by winding it, and for performing other functions. These are some of the examples of using an existing object for other purposes.

Figure 4.3 shows clips, which are commonly used to organize documents. Such a clip may be used for purposes other than merely filing documents. You may straighten it up and poke your friend at his arm or use it like a wire. These are two examples of how an object may provide various functions aside from its original purpose.

The next item is a light bulb (Fig. 4.4). The primary goal of a light bulb is to illuminate dark areas as illustrated in the left figure. Light bulbs used in traffic lights, however, do not aim to brighten the street. They indicate traffic signals with distinct colors. Their original purpose is not applied in this case. The F1 rule is applied here.

Fig. 4.3 F1: function

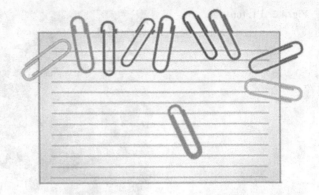

Fig. 4.4 F1: function

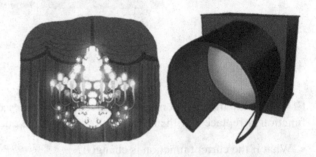

Fig. 4.5 F1: function

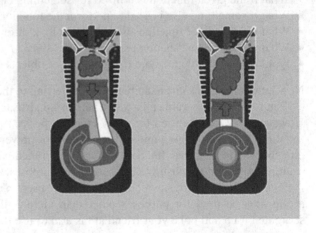

Figure 4.5 shows the engine of a car. Fuel passes into the upper part of a car engine and explodes. The explosion prompts the piston to move up and down. The vertical momentum then is converted to a rotational motion by the crank at the bottom. The same amount of force is transformed from a lineal movement to a rotational movement, which incites the wheels to rotate and the car to move forward.

Fig. 4.6 F2: fertileness

Fig. 4.7 F2: fertileness

4.3 F2: Fertileness

The second rule applied to the field axis is F2 or Fertileness. Fertileness indicates wealth of materials, and the needlessness to stick to a given material.

- What if another material is used instead?
- What if two or more materials are combined?
- What if the existing material is no longer used?
- Could another material be possibly used additionally?

Let us examine some pictures in this regard. Figure 4.6 shows two TV sets. In the past, TV screens used to be made of cathode-ray tubes (CRT). TV sets were big because of such CRTs. As LCD or LED units replaced these tubes, TV sets became light and slim, which was another innovation achieved by changing a material or component.

Figure 4.7 shows wires used for communication services. Copper was commonly used for telephone wire before, but it was replaced with optical fiber, which resulted in various advantages in terms of lightness and performance. This is another example of changing a component and gaining better results.

Fig. 4.8 F2: fertileness

Figure 4.8 is about CFC (Chlorofluorocarbon, Freon), which is used for refrigerators and air conditioners. As CFC involves no chemical reaction, it has been developed and used as a gas refrigerant; that is, a gas that facilitates heat emission from refrigerators and air conditioners. That this gas might cause ozone layer depletion and natural environment damage were recognized, however. While the ozone layer that blocks ultraviolet rays from the universe is destroyed, excessive ultraviolet rays cause harm to living creatures on earth. Thus, mankind has to reduce the usage of CFC and ultimately, develop another material to replace it. F2 is involved in this case and a new material is sought as well.

4.4 F3: Fusion

The third rule on the field axis is F3 or Fusion. In application of this rule, existing functions may be combined or linked to functions in other fields, or specific performances can be combined as the following questions are raised:

- What if a function is combined with other functions?
- What would it result to if existing A and B functions were combined?
- What functions must be combined for the specific function that I need?
- What if various functions work in parallel?

The tablet PC shown in Fig. 4.9 has no keyboard. This is a fusion of a keyboard and a display. In general, a computer has a keyboard and a screen, but this device includes a keyboard on the display in a form of a touch screen, which is an example of functional fusion.

The smart phone in Fig. 4.10 is another example. This combines the functions of existing cell phones, cameras and those of computers. Smart TVs that are used recently also include some of computers' features.

Figure 4.11 shows a digital camera. This camera has photographic and telecommunication functions. Once you take pictures, you can send them via SMS or the

Fig. 4.9 F2: fertileness

Fig. 4.10 F3: fusion

Fig. 4.11 F3: fusion

Internet. This is a good example of what electronic products can transform to if given unconventional functions.

Figure 4.12 shows a copy machine. This unit not only has the features of an existing copier but also those of a scanner, such as a scanning screen and digital image saver, in addition to printing out contents from computer files. The functions of a copier, a scanner, and a printer are all combined in one unit.

Fig. 4.12 F3: fusion

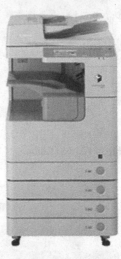

Fig. 4.13 F3: fusion

Figure 4.13 shows a TV monitor that is used in common households. A wide screen would be good while you watch TV, but otherwise, you may not like the black screen that occupies a large portion of the room. Think of ways to utilize this for other purposes. The author is thinking of how it would be like if the TV set displays images while it functions as a mirror. This is an idea in application of the rule of Fusion.

4.5 Hyundai Card Benchmarking

Benchmarking is a way to learn from things. Companies or schools may compare themselves with other institutions to learn the latter's advantages. Such benchmarking involves organizations in the same field, but Hyundai card has achieved great success by thinking outside the box.

Hyundai card currently holds 10.9 % of the market, ranking the first or second in Korea. About 10 years ago, however, this company was at the very bottom of the field, accounting for 1.8 % in the market. Such a rapid growth resulted from benchmarking other fields.

Hyundai card has been known to take steps that other credit card companies would never seize and this trait contributed to its growth. This company developed unique marketing strategies such as using prepaying point service and releasing ads totally different from those of other credit card companies. A design concept was also introduced by it to credit cards to alter the image of hard plastic and represent friendliness.

Further, it held large-scale music concerts and tennis matches, which were never expected among credit card companies. This company introduced such unique strategies with regard to corporate culture, advertisement, marketing, and personnel management that common card companies would not think of.

What was the catalyst for Hyundai card's extraordinary behavior? It was benchmarking of other fields. In general, benchmarking involves a comparative analysis of other agencies or organizations in the same field, but Hyundai card studied entities in other fields.

It analyzed companies in fields that seemed to have nothing in common with the credit card business such as hotels, art halls, newspapers, museums, etc., and compared the advantages and disadvantages of these entities with theirs. Through surprising and unfamiliar concepts introduced in the card business, they were able to attract customers.

It is said that there is nothing new in this world, but combining existing things in various ways can lead to a sort of creation. To this end, you need to observe other things closely. Look into other fields, turning your eyes from your own work, major, and company. There may be some clash of opinions among people with different majors at school or work but insisting one's own field leads to narrow-mindedness. This is why some say that you need to become a "T-type human," a person with in-depth expertise in a certain field and a wide range of knowledge at the same time.

4.6 The Discipline of Division Convergence

The term "convergence" reminds me of a personal anecdote. It is the story of the founding of the Department of Bio and Brain Engineering of our university for which I work.

In 2001, I had been working at a computer science department for 17 years. Back then, IT (Information Technology) was showing the strongest growth. I was wondering what industry would rise after IT. After discussing with many people, I concluded that the bio technology industry would be a major industry of the 21st century. As a professor in the field of computers, I thought about ways to contribute to bio technology. I thought that opening up a field of convergence for bio technology and information technology would create considerable contributions.

I suggested to the school office the establishment of a department dedicated to the convergence of bio technology and information technology. In this new department, the teaching staff from such areas as computer, electronics, biology, medical science, and mechanical engineering would study and teach together. I received a USD 30 million fund from Chairman Moon Soul Chung of Mirae Corporation for the department's foundation.

But the school reacted negatively. Would it be really possible to combine biology and electronics into one curriculum? Would the graduates be able to get a job after completing the course of neither biology nor computer science? Almost all were skeptical.

For some time, fields of research and education such as physics, chemistry, mathematics, electronics, mechanical engineering, business management, and so forth, have been regarded as confined disciplines. Many thought that there could be no branch of learning that combined two or more fields at the same time. They said that there would be no occupation for graduates from such departments.

After many setbacks, the department was finally established. Fellow professors who joined me in the department tried to create a vision to interested students and encouraged them to choose the newly established department. It was not an easy matter to bet your future on a department with no history.

Meanwhile, foreign universities in the U.S. such as Stanford University and MIT established a department of bio engineering, and many similar departments started to be founded overseas as brain research made progress. About 10 years later, the configuration of disciplines and industries changed a lot. Similar departments have been established elsewhere in the world. The subject of convergence has become a trend in academic circles. It became an established theory that convergence produced fresh ideas. Large scale companies started to invest in the bio equipment industry.

Graduates from the bio and brain engineering departments became so precious that they began to have a variety of choices for employment. They looked ahead and welcomed a new field rather than join others in long-existent and seemingly secure courses. As the number of universities that launched departments combining bio and electronics increased, available positions for professors increased accordingly.

4.7 Misjudgement in History That Resulted from Obsession

Germany and the former Soviet Union concluded the mutual nonaggression pact on August 23, 1939. Shortly after that, however, Germany invaded Poland from the west on September 1, and the Soviet Union attacked it from the east on September 17. These two ended up dividing the country into two parts under their respective rule. They abided by the agreement by sharing and occupying the spoils. They went further by concluding the Germany Soviet Treaty of Friendship on September 29 to strengthen their ties.

Giving reassurance to the Soviet Union by concluding the treaty, Hitler and Germany focused on the battlefront to the west. From April to June, 1940, he advanced from the Netherlands to Luxembourg and to France.

Germany was in an advantageous position to conquer the European continent once it could defeat the Soviet Union to the east. Aflame with the desire, Hitler started to prepare for attacking the Soviet Union.

Stalin in the Soviet Union did not even imagine a German attack, on the other hand. Various information sources reported the possibility of German invasion, but he ignored them. Stalin confuted to their faces generals who insisted to prepare for the attack based on specific information. The Soviet Union ended up facing German armies without even starting to prepare for the war. While hundreds of thousands of German soldiers were ready for the war, the Soviet Union were not prepared at all.

Finally, Germany pursued an all-angle attack to the western forward edge of the Soviet Union on June 22, 1941. Hitler boasted that his army would occupy Moscow in the coming fall, and it seemed as though his word would become reality. Although Stalin heard reports about the German invasion, he underestimated it, thinking that it was mere local war. With no defense, there was nothing that the armies of Stalin could do but concede. Soviet armies scattered or surrendered while the German army continued to move forward up to the point of almost entering Moscow in September.

But Hitler made a misjudgement this time. He thought that it would be necessary to secure agricultural products and natural resources to continue the war; thus, he ordered his main-force units to focus on Ukraine instead of Moscow although his generals insisted to advance on Moscow since they had to finish the war before winter came. Hitler rebuked the generals who disagreed with his opinion and told them to carry out the order.

In the meantime, Moscow bought time while Hitler tarried in Ukraine. In October, news spread that the German army was about to occupy Moscow. The German flying corps frequently conducted air raids. The Soviet government chose Kuybyshev as its provisional capital and planned to withdraw its army from Moscow. In mid-October, special evacuation trains started to run. Diplomats stationed at Moscow and high officials moved to the provisional capital where the 40-m deep underground bunker for Stalin was being constructed.

On October 16, the terror over Moscow reached its peak. The streets were flooded with refugees from the city which fell into anarchy. Questions on whether Stalin would leave Moscow were raised. On October 18, Stalin arrived at Kursk station where a special train for him was waiting. Stalin approached the train but suddenly walked up and down the platform. He left the station instead and declared that he would not leave Moscow.

On October 19, Stalin declared martial law and tried to maintain order in the city. He ordered for an army parade on November 7, the memorial day of the

Bolshevik Revolution. It posed as a highly dangerous event as the German air force continued with their bombing raids. Stalin projected a sense of security through the army parade and showed his will to defend Moscow. The German army became rapidly weakened in the face of the extremely cold climate while the morale of the Soviet army was boosted. The war was not going in the German army's favor. Hitler ended up admitting final defeat in April 1942. As the invincible German army went downhill, World War II came to its end.

The two dictators made a serious misjudgement during the war. Stalin ignored a lot of information on Germany's invasion and almost gave up Moscow once the war broke out. Hitler underestimated the intensity of the Soviet Union's climate. He took no heed of the generals' idea to occupy Moscow before the cold came, wasting time in Ukraine. Their misjudgements resulted from their blind adherence to one information while ignoring other possibilities.

4.8 Welcoming, Tolerating, and Amalgamating Things

As shown in the examples earlier, it is difficult to come up with fresh ideas once you fasten your eyes on a certain field and fall into the trap of fixed ideas. It is necessary to move your line of thought according to the direction of a navigation. When two different fields run into each other, a fresh idea may be born.

Today, automobiles have become a product of convergence to the point that it has become difficult to tell if they are machines or electronic appliances. State-of-the-art automobiles are a good example of innovation achieved by the convergence of fields.

Thinking in an integrated manner by placing objects in other fields is of significance not only in coming up with new ideas but also in terms of wisdom in life. We are likely to stick to and insist a certain view. This is because we are absorbed in reality and unable to see the things around.

Looking at other "fields" means to think with an open mind. It is considering others' position or trying to understand others in their positions. The old saying "*Put yourself in others' shoes*" applies in this case.

Opening to the world means embracing other ideas and combining them. There is no organization or nation that prospers without opening itself. Openness, tolerance, convergence, and integration are different words but stand for one principle. This may remind you of *Day of Empire* written by Amy Chua which was mentioned earlier. The principal elements to become a world power are openness and tolerance. Nations that opened themselves to and embraced foreign cultures prospered, but those that closed themselves fell in history.

Zhou Enlai, who helped build modern China along with Mao Tse-tung, was a master of negotiation. He would say when he learned that his idea was different from another, "I like it when we have things in common and yet possess different ideas at the same time. Let us proceed with things that we both agree on with our differences respected."

Richard Carlson says in his book *Don't Sweat the Small Stuff,* "Do not struggle with minor matters for only then can you learn how to embrace a lot of new things."

4.9 Summary

1. When using F1 or Function rule, the following points may help:
- Consider a cup with electronic features.
- What if it indicates the temperature?
- What if the color changes depending on the temperature?
- What if the cup is edible?
- Could the paper be made of materials other than natural pulp?

2. In application of F2 or Fertileness rule, consider the following questions:
- What if another material is used instead?
- What if two or more materials are combined?
- What if the existing material is no longer used?
- Could another material be possibly used additionally?

3. In application of F3 or Fusion rule, consider the following questions:
- What if a function is combined with other functions?
- What would it result to if existing functions A and B were combined?
- What functions need to be combined for the specific function that I need?
- What if various functions work in parallel?

4. The saying *Put yourself in others' shoes* is in line with the idea of applying things in other fields.
5. The principle of tolerance is in relation to the idea of tolerating and understanding other fields.

4.10 Exercise

1. Find and discuss cases to which both F1 and F2 can be applied.
2. Find and discuss cases to which both F2 and F3 can be applied.
3. What questions may be applied to designing a mountain bicycle?
4. When designing a flying automobile, which rule can be applied to modern day automobiles?
5. If you wish to increase a current peak speed of 200–300 km/h, which question may be applied?
6. Discuss the secrets of Steve Jobs' innovative achievements.

7. Why do people think of paper cups when it comes to disposible cups?
8. What questions may you ask when designing products that combine the functions of a smart phone and TV?
9. What questions may you ask when factorizing in mathematics?
10. What questions may you ask when making things move faster?

Chapter 5
Traveling on a Two-Dimensional Plane

Abstract Without any external stimulus to brain, a person would stick to certain moment, space or field and the mind would remain in its current situation. Traveling on axis of time, space and field can make the brain free from the reality. In this chapter, a method of traveling on two axes simultaneously is proposed, that is traveling on two-dimensional plane. For example, traveling on time-space plane helps us to consider the time and space factors together. On each axis, there are three rules, and thus combination of two axes gives nine combined rule of making questions.

Keywords Creativity • Time • Space • Field • Extension of thought • Traveling• Innovation • Frame • Think

As mentioned earlier, there are three axes we may explore: the axis of time, the axis of space, and the axis of field. Two axes can make a plane, and thus this chapter addresses questions that may arise on a plane where two axes cross. Spatial environments change over time as anything does. Let us travel and seek ideas while imagining changes in space and field over time.

5.1 A Plane That Involves Two Axes

Bring the disposable cup stated above back to mind this time on a time-space plane. How would disposable cups change in the future's space? A variety of ideas may come up as you freely move this given question on the plane.

- What forms and materials may be used for paper cups 30–40 years from now?
- In which shape would cups be used 20 years from now in China?
- What would cups be like 10 years from now in Islamic cultures?
- In which shape would cups be used 10 years from now in Siberia?

K. H. Lee, *Three Dimensional Creativity*, KAIST Research Series,
DOI: 10.1007/978-94-017-8804-5_5, © Springer Science+Business Media Dordrecht 2014

- What type of disposable cups would be needed 20 years from now in desert regions?
- What type of cups would be used if global warming continued?
- What type of beverages would people enjoy drinking on a boat 20 years from now?

Let us think of disposable cups on a plane of time and field this time. What questions might arise?

- In 10–20 years from now, what changes would there be in the materials and functions?
- What type of beverages would people enjoy drinking on a boat 20 years from now?
- Would the convenient functions of disposable cups be able to apply somewhere else?
- Could artificial pulps be used in the future instead of the natural pulps commonly used today?
- What characteristics would cups made of natural pulps have?
- What characteristics would paper made of natural pulps have?

Various questions as above have already arisen. Thinking of anything on the axis of time produces a new environment and fresh ideas accordingly. It may be difficult to come up with such ideas without the frame of thought suggested here.

5.2 Rules in Combination of Time and Space

Three sub-rules—T1 (transpose), T2 (tempo), and T3 (translation)—were made on the axis of time, and another three—S1 (Shape), S2 (Site), S3 (Size)—on the axis of space in the previous chapter. Since there are three rules under each axis of time and space, a combination of these rules results in nine rules. This section examines how these nine rules may apply (Fig. 5.1).

Figure 5.2 shows a child who is putting Lego together. There may be a certain order of putting Lego blocks together, and this order decides the formation. Changing the formation when playing with Lego pieces changes the order of construction, and the result also changes. The combination of T1 on the axis of time, which is the rule of transpose, and S1 on the axis of space, which is the rule of shape, creates the idea of putting Lego pieces together.

Figure 5.3 shows a motorcycle running on curves at a high speed. It seems that the motorcycle is lying on the side completely but it is moving forward. As it accelerates, its center of gravity lowers and thus it can move forward without falling down. This is an example of simultaneously applying T2 (tempo) on the axis of time, which is a rule of speed, and the rule of S2 (site) to the center of gravity in order for the motorcycle to run on the curves at a high speed.

Figure 5.4 shows a cubic puzzle. It has 6 different faces, each of which is supposed to have blocks with the same color. To reach this solution, you have to turn

Fig. 5.1 Time–space plan

S1: Shape,
S2: Site,
S3: Size

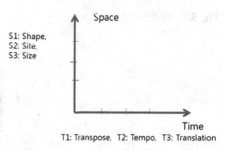

Space

Time

T1: Transpose, T2: Tempo, T3: Translation

Fig. 5.2 T1 (transpose) + S1
(shape)

Fig. 5.3 T2 (tempo) + S2
(site)

Fig. 5.4 T1 (transpose) + S2
(site)

the cubic blocks in a certain order of movement. The order may vary depending on the location of the cubic color. This is another example of combining T1 on the axis of time, which is the rule of transpose, and S2 on the axis of space, the rule of relocation, to put a cubic puzzle to its original state.

Fig. 5.5 T1 (transpose) + S2
(site) + S3 (size)

Fig. 5.6 T2 (tempo) + S1
(shape)

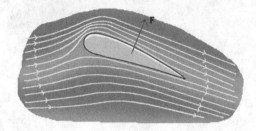

Figure 5.5 shows a drawing of a semiconductor chip. An electronic circuit is inserted into this tiny chip and electronic elements are connected for this circuit in the designated order. Electronic elements would be connected usually on a plane circuit, but in this case, the electronic circuit is arranged on various layers since a single-layer circuit would occupy a larger space. The order of arranging electronic elements is adjusted accordingly, they are arranged on various layers, and thus the general size is reduced. This is an example of applying T1 (transpose) and S2 (site) to cause changes in S3 (size).

Figure 5.6 illustrates the basic principle of the flight of an airplane in the sky. When seen from the side, airplane wings seem to be curved asymmetrically. As the airplane moves forward at a high speed, the speed of airflow at the lower part of the wings becomes different from that at the upper part. The pressure at the upper part is reduced since the speed of airflow increases. As the pressure lessens, the wings are pushed up according to Bernoulli's principle which pertains to the lift force on the wings. This is what's behind the principle of an airplane's flight—the combination of T2 and S1 results in the principle of the flight.

Figure 5.7 shows a spray of substances, which is a similar example. As mentioned above, Bernoulli's principle is applied as the inlet is quite small so that the pressure of the liquid coming out of the container at a high speed goes down and the liquid is scattered in the air. T2 (tempo) and S1 (shape) are applied in this case.

Fig. 5.7   T2 (tempo) + S1
(shape)

Fig. 5.8   T2 (tempo) + S1
(shape)

This is the case when you blow your nose (Fig. 5.8). It is difficult to blow your nose without using your hands. Hold your nostril with your hands to make it smaller. Change the shape of the nostril and then blow air into it at a high speed. With the pressure within the nostril lowered, the nasal discharge inside comes out. This is another example of applying T2 and S1.

A series of mathematic expressions are presented in Fig. 5.9. Look at Expression 1. This shows the relation between x and y, but it is not easy to notice instantly what relation x and y have. In application of the rule of transpose, however, change the order (location) to get Expression 2. Divide $\sqrt{2}$ at the left side to both sides to get Expression 3. Since $\sqrt{2}$ is part of the denominator in the right side, multiply the numerator and denominator with $\sqrt{2}$ to move it to the numerator. Upon that point you will get Expression 5. Expression 5 makes it easy to

Fig. 5.9 T1 (transpose) + S2
(site)

$$\sqrt{2}\,y^2 = -\sqrt{2}\,x^2 + 2 \qquad (1)$$

$$\sqrt{2}\,y^2 + \sqrt{2}\,x^2 = 2 \qquad (2)$$

$$y^2 + x^2 = \frac{2}{\sqrt{2}} \qquad (3)$$

$$y^2 + x^2 = \frac{2\sqrt{2}}{2} \qquad (4)$$

$$y^2 + x^2 = \sqrt{2} \qquad (5)$$

Fig. 5.10 T1
(transpose) + S2 (site)

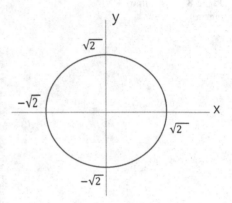

notice that the relation between x and y produces a circle whose radius is $\sqrt{2}$. This is an example of combining the two rules: T1, the rule of transpose, and S2, the rule of site (Fig. 5.10).

5.3 Rules in Combination of Time and Field

This section addresses the questions that may arise on a plane where time and field cross. The three rules of time—T (transpose), T2 (tempo), and T3 (translation)— and the three rules of field—F1 (function), F2 (fertileness), and F3 (fusion)—are defined. The combination of the two axes produces $3 \times 3 = 9$ rules. Questions may turn up when you apply these rules.

You may well know of Kodak Film (Fig. 5.11). This company in the U.S. was a traditional high-tech business for 133 years, but lately it went bankrupt. Why? Kodak made a wrong decision on the plane of time and field. Kodak grew into a big company as it developed traditional optical films. In 1975, this business

Fig. 5.11 T1 (transpose) and
F1 (function)

Fig. 5.12 T1 (transpose) and
F1 (function)

developed the original technology of digital camera independently. It made a
serious mistake, however, when it neglected the potential of digital camera and
focused on producing optical films. In contrast, Fuji Film, which was second to
Kodak in the optical film industry, made a different decision.

Fuji as well had a long history after its foundation in 1934. As the digital film
technology developed, it noticed that the optical film market was going to decline.
Based on its original optical technology, Fuji moved its momentum of business to
another sector in close relation to the major industry, and it had remained as a busi
ness with internal stability since it expended its business areas to medication and
cosmetics. A comparison of Kodak and Fuji illustrates the importance of predic-
tion; that is, emphasizing the pre-action of transpose. Looking ahead on the axis
of time and deciding the next major business sector is the core of future-oriented
strategies. This is another example of combining T1 (transpose) and F1 (function).

Recently, Sony in Japan has been facing difficulties as well (Fig. 5.12). This
company has neglected digital TV technology that emerged from the late 1900s
to the early 2000s. It failed to convert its TV technology to digital type in a
timely manner. Having grown as a world business by developing innovations like
Walkman, Sony mistakenly selected a major sector on the axis of time, and is now
facing serious difficulties. This is an example of a wrong answer to questions on
the plane of T1 and F1.

Figure 5.13 shows a touch screen of smart phone. The touch screen is used to
input letters and symbols in order, and the order is of importance. Changing the
order leads to different functions. Different orders result in new functions, and
orders are changed on purpose for different functions. This is an example of combining T1 and F1.

Figure 5.14 shows a medicine called Botox. Botox was developed in the 1980s
originally to alleviate tics. It was observed, unexpectedly, that using this medicine
to relax the muscles resulted to swelling and wrinkle treatment. The primary function changed. This is an example of changing F1 through T3 (translation) via a
different interpretation of the medicine. In the area of new medicine development,
the elements and functions of existing medicines are examined to utilize them for
different functions.

Fig. 5.13 T1
(transpose) + F1 (function)

Fig. 5.14 T3
(translation) + F1 (function)

5.4 Rules in Combination of Space and Field

This section addresses the area where space and field coexist. Here, three rules of space—S1 (Shape), S2 (Site), and S3 (Size)—are applied. These rules meet the rules of field–F1 (function), F2 (fertileness), and F3 (fusion)—and produce nice sub-rules. Let us consider the following examples.

Fig. 5.15 S1 (shape) + F1 (function)

Fig. 5.16 S1 (shape) + F1 (function)

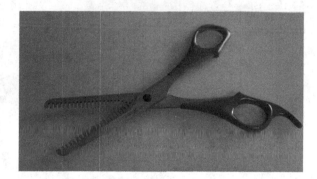

Figure 5.15 shows a spoon with a fork-like edge so that it can function as both a spoon and a fork. This demonstrates a combination of functions with the shape of the spoon edge changed to combine the purposes of two utensils.

Figure 5.16 shows haircutting scissors. Usually, scissors consist of two sharp blades, but the edges of the scissors in the picture are designed to look like saw blades in order to have hairs cut naturally and not in the same length. The product goes beyond the fixed idea that scissors cut things uniformly. This is an example of applying S1 (shape) and F1 (function).

Figure 5.17 shows a humidifier that adds humidity in the air. Developed by Prof. S. M. Bae at KAIST, this unit looks like a flowerpot, different from existing humidifier models. Further, it is made of paper. The shape was changed (S1), and a different material was used (F2) to create new functions (F1).

Figure 5.18 shows a flexible display. Existing flat displays at one point were of LCD type. Since LCD uses a screen backlight, it didn't allow for a curved screen. Lately, the technology of using LED materials emerged. Since LED consists of luminous material and involves no need for backlight, its form may be changed as desired. The figure shows LED TV and smart phone models with a curved screen. Curved TV screens enhance a sense of three-dimensional structure, and bendable smart phones are easy to carry. This is an example of combining S1 (shape) and F2 (fertileness).

Fig. 5.17 S1 (shape) + F1
(function) + F2 (fertileness)

Fig. 5.18 S1 (shape) and F2
(fertileness)

Figure 5.19 shows a high-grade bag. This sort of name-brand bags costs tens of thousands of dollars. Although its original function was just to carry belongings in it, its main function has changed to showing social status. Its major function changed as its design became better. This is an example of combining S1 and F1.

5.5 Innovation of Steve Jobs

There is no completely new idea under the sky. Combining existing facts or knowledge makes products innovative. The research conducted by Prof. Clayton Christensen at the School of Business Management at Harvard University in the U.S. also supports this. The research involved interviews with 3,000 executives in companies for six years. This research indicates that "creativity is a combination of existing facts or knowledge rather than creating a totally new thing." If this is the case, we need to gain fundamental knowledge as much as possible. This is why those who have a wealth of experience and knowledge are likely to produce new ideas.

Regarding Steve Jobs, Camine Gallo stated in his book entitled *Innovation Secrets of Steve Jobs,* the distinguishing characteristic of Steve Jobs is his "integrated thinking." Apple, the company he founded was a technology–based

Fig. 5.19 S1 (shape) + F1 (function)

manufacturing company. However, he did not stick to technology. He instead combined it with other sectors and produced fresh ideas.

One of his strong points was that he would connect things that did not seem to have things in common and created original ideas. His combined knowledge and experience became good materials to produce more combinations. New experience produced new combinations, which stimulated the brain and caused it to produce new ideas. Steve Jobs was interested a lot in Indian meditation, music, calligraphy, and humanities. He would combine his knowledge on these areas with technology to create original things sequentially on a time horizon.

5.6 Science Education That Helps Learners Understand Why They Learn

Although I have continued learning up to the point of taking a graduate course in the university, I am still ignorant of the meaning and usage of some of the mathematic functions that I've learned from school. I do not know how they could apply to real life. I had to memorize some functions just to prepare for an exam even though I did not understand their principles. After the exam, I simply forgot those lessons. This is something to be overcome in the education of students in mathematics and science.

The goal of science education is to help understand the principles of nature, to develop curiosity, and to cultivate potential capabilities. If you do not understand the meaning and usage of things you learn, it would be difficult to understand their principle and to show interest in them.

In the real world, everything is connected. In a classroom, however, disciplines such as mathematics, physics, chemistry, and biology are taught separately and recognized as irrelevant. It is necessary to sometimes link objects that seem to be of different categories. Afterwards you may cry out, "The things I learned in chemistry class the other day are related to physics like this!"

The "frame" of teaching mathematics and science at school needs to be changed. It is not that you have to learn since it will be necessary later, but that you have to understand why the learning material is necessary in designing things actually used in daily life.

It would be desirable to add a class for "science technology" to middle and high school curriculums. There is no need to take much time. It would be enough if there is an extracurricular activity that introduces cutting-edge products of IT (Information Technology), BT (Bio Technology), NT (Nano Technology), ET (Environment Technology), and so forth.

It would be effective if an object is shown during the class. Products reflect cutting-edge technology, and basic knowledge of mathematics and science is also beneath the surface. When the principles in a textbook are presented this way, students will readily understand how the things they are learning are used in cutting-edge products. Mathematics, physics, and chemistry are fused in a product. Products help students understand how different subjects are connected to each other.

A cell phone, for example, may be used to explain the basic principle of wireless telecommunication. You may explain the electronic circuits and software in it, and the knowledge in mathematics, physics, and chemistry involved to make the phone suitable to middle and high school students. In an automobile class, the basic technology to enable a car to run fast may be explained with the related contents in a mathematics and physics textbook.

The way of learning materials in a textbook with a related product presented is called RSP (Reverse Science from Product). This is in line with the direction of STEM (Science, Technology, Engineering, and Mathematics) or STEAM (Science, Technology, Engineering, Art, and Mathematics).

It is often said that instructors need to arouse interest in science among students and motivate them to study. The actual usage of science, however, has not been widely illustrated in classes. Science technology education that connects products used in real life and contents in textbooks will be one solution in this regard.

5.7 Summary

1. The combination of three rules on the axis of time (T1, T2, T3) and the three rules on the axis of space (S1, S2, S3) results in 9 rules (e.g., T1 + S1, T2 + S1, T3 + S2, etc.).
2. The combination of three rules on the axis of time (T1, T2, T3) and the three rules on the axis of field (F1, F2, F3) results in 9 rules (e.g., T1 + F1, T2 + F2, T2 + F3, etc.).
3. The combination of three rules on the axis of space (S1, S2, S3) and the three rules of fields (F1, F2, F3) results in 9 rules (e.g., S1 + F1, S2 + F2, S3 + F2, etc.).

4. The findings in Professor Clayton Christensen's research indicates that "A new thing is created when existing facts or knowledge are combined."
5. The secret of Steve Jobs' innovation is integration.
6. Science that you learn while seeing products used in real life is called RSP (Reverse Science from Product). RSP presents the cases where science is practically utilized and various disciplines are combined within one product.
7. STEM (Science, Technology, Engineering, and Mathematics) or STEAM (Science, Technology, Engineering, Art, and Mathematics) is a suggestion to teach the students in an integrated way.

5.8 Exercise

1. Apply T1 + S1 to a bike and make questions.
2. Apply T2 + F2 to an automobile and make questions.
3. Apply T1 + S1 to TV and make questions.
4. Apply T3 + S1 to putting Lego pieces together and make questions.
5. Think of questions about TV to make a transparent one.
6. Think of questions that may arise to make a foldable keyboard. What kinds of questions can be given?
7. Think of questions that may arise to make a foldable display that can be carried in a pocket.
8. Think of questions that may arise to make an unmanned airplane.
9. What if S1 + F2 is applied to a balloon?
10. Think of questions about dolls to make a doll that plays the piano.

Chapter 6
Traveling in a Three-Dimensional World

Abstract Almost everything in the world are related to three basic elements, and thus questions about them can touch almost all the important aspects of given problems. Combining these three questions results in forming a three-dimensional world, and you can free yourself from reality as you move on the three axes in the three-dimensional world. In such a free state, the brain produces a wealth of new ideas. How could we make ourselves ask questions when alone? It is necessary to make it a habit to ask the three questions. Repeating makes neural circuits in our brain which corresponds to the habit. As the process of questioning is repeated, it will become a habit, and this ingrained tendency will help us to become a creative person.

Keywords Creativity • Three-dimension • Traveling • Time • Space • Field• Neural circuit • Brain • Neuron • Habit

Traveling by thought through the three navigation systems of time, space, and field was suggested earlier. In each case, the rules T1, T2, and T3 were given on the axis of time; S1, S2, and S3 on the axis of space; and F1, F2, and F3 on the axis of field. You may make questions when dealing with each rule while traveling on the three axis.

Each suggested navigation corresponds to a sort of coordinate axis. When these three axes are combined, a TSF three-dimensional world is formed as shown in the figure. The sets of three rules defined on each axis can be combined in this three-dimensional world. The total number of rules is 27 ($3 \times 3 \times 3$). Let us look at the examples in Fig. 6.1.

K. H. Lee, *Three Dimensional Creativity*, KAIST Research Series,
DOI: 10.1007/978-94-017-8804-5_6, © Springer Science+Business Media Dordrecht 2014

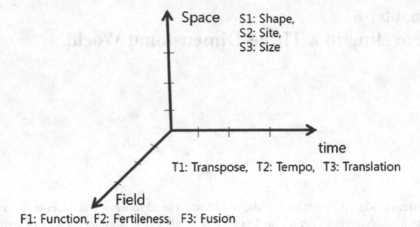

Space   S1: Shape,
         S2: Site,
         S3: Size

time

T1: Transpose, T2: Tempo, T3: Translation

Field

F1: Function, F2: Fertileness, F3: Fusion

Fig. 6.1 3 Dimension (time–space–field)

Fig. 6.2 T2 (tempo) + S1 (shape) + S3 (size) + F1 (function)

6.1 Ideas in the Three-Dimensional World

Figure 6.2 shows a portable bicycle. In general, bicycles are not portable but they are used to travel fast. Their weight and size are not appropriate for carrying. The model in the figure, however, is different in shape although its speed might be relatively slow. Its weight and size are also smaller for different functions. This is a product to which T2 (tempo), S1 (shape), S3 (size), and F1 (function) are applied.

Fig. 6.3 T2 (tempo) + S1 (shape) + F1 (function)

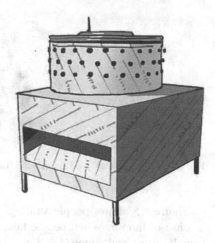

Fig.6.4 T2 (tempo) +S1(shape) + F1(function)

Figure 6.3 may look somewhat unfamiliar. This is a machine that slaughters chickens and removes their feathers before they are cooked. Removing feathers was used to be done with hands, but now this process is automatically done if you put dead chickens into this machine and turn it on. This machine utilizes a principle that may apply to spin-dryers. It spins the tub as fast as a spin-dryer. The embossed rubbers within the tub rub against chicken feathers so that they are pulled up. The shape of the inner section is different from that of a spin-dryer so as to change its function as well. This is an example of a product made from the combination of the T2 (tempo), S1 (shape), and F1 (function) rules.

Figure 6.4 shows a chestnut-peeling machine. Its function is similar to the drying function of a washing machine. There are small blades within the rotating tub. If it rotates at a high speed with chestnuts in it, the chestnuts are peeled. Increasing the rotating speed and changing the shape leads to a product with new functions. This is an example of combining the T2 (tempo), S1 (shape), and F1 (function) rules.

Fig. 6.5   T1 (transpose) + S3 (size) + F3 (fusion)

Figure 6.5 shows people smiling. I have thought a lot about how to develop a sense of humor. When people talk, changing the order of a story may make it sound absurd and funny. Such a change of order in a conversation corresponds to T1 (transpose). It may be difficult for an ordinary story to be funny. Thus, the story may be exaggerated or expanded in size (S3). This can be combined with a context which is different from that in the existing story (a combination with F3). Thus, humor can be spoken of as an application of the T1 (transpose), S3 (size), and F3 (fusion) rules.

6.2   The Three-Dimensional Future-Oriented Strategies of K-Pop Music

K-pop or Korean popular music, is becoming globally powerful. Some of the performances of K-pop acts made the young people in Paris and New York, both centers of world culture, absolutely enthusiastic. Recently, Girl's Generation, a K-pop girl group, has appeared in talk shows of well-known broadcasting channels in the US such as CBS, ABC, and NBC, and attracted popularity. An original strategy that goes beyond the border of space and field made Korea overcome the limit of its national boundary (Fig. 6.6).

SM Entertainment, the company taking the lead of K-pop waves, promoted its plans in a unique manner right from beginning. It analyzed world music trends for its planning, defining public performance as something that entertains not only ears but multiple senses. It tried to maximize visual movements such as dance and performance in addition to beautiful melodies. Going beyond the existing genre of music, it integrated dance, visual elements, and stage performance to create a systematic stage art.

SM Entertainment initiated an exchange of talent with world musicians more than a decade ago right from the stage of planning. To advance in the world, there had to be something more than just music. It continued to establish a network to reach world-class composers, dancers, album distributors, and performance experts, among others. In addition, it tried to study and embrace world favorites.

Fig. 6.6 T2(transpose) + S1(site) + F1(fusion)

Although its songs were sung by Korean singers, it did not insist Korean styles. It joined world composer groups to ask for their compositions or participation in projects. "Pinocchio," a song of one girl group called f(x), was composed by an American composer, Alex Cantrell. "Tell Me Your Wish", sung by Girls' Generation, was produced jointly by musicians in Norway and the U.K. In many of such musical works, fast beats from Westerners were combined with Korean melodies.

The dance was simple but it involved big and powerful motions. Simple motions were repeatedly performed by a group of singers as part of their dance. It seemed as though a group of cranes were dancing. The motions were so easy that anyone could join along. More than 10 dance experts from abroad participated for the choreography's construction. Nick Bass from the US and Rino Nakasone from Japan are two of them.

Rino Nakasone is well-known for his dance compositions for Michael Jackson. Rino Nakasone participated in dance compositions for "You Are Like an Air" and "Juliet" by SHINee and "Tell Me Your Wish" by Girls' Generation. Such dance compositions as "You Are Too Pretty" by SHINee and "Chu" of f(x) are works of Nakasone too.

Rino Nakasone put special focus on the simple motions. The jegi kicking and the arrow fitting motions in "Tell Me Your Wish" by Girls' Generation, for example, are quite simple. The hip dance in "Mister" by KARA and the repeated motions in "Nobody" by Wonder Girls are among other examples.

The motions need to be simple for the world to understand and follow. The points that would attract the world need to be found and emphasized with the rest left out.

Selecting singers went beyond the boundary of Korea as well. Nichkhun of 2PM is one of the most successful foreign singers in Korea. Imported from Los Angeles, USA in 2006, Nichkhun who was originally from Thailand started his training in JYP Entertainment and became a member of the successful dance group a few years later.

Time (future-oriented strategies)	Planning 10 years ahead
Space (globalization)	Composer: Alex Cantrell
	Dance: Rino Nakasone
Field (integration)	Composite arts of music, dance, and stage performance

[K-pop's time-space-field journey]

Singers aiming at overseas markets from the beginning are trained in music and dance. KINO is a male group aimed at markets in Japan right from the beginning. They learned Japanese cultures and language in Japan. They made their stage debut in Japan and sung songs in Japanese as well.

Everything about this group was planned from a global perspective. They did not insist a Korean style, but tried to go beyond its border or "space". They did not stick only to music. Rather, various "fields" from music to dance to "wire way" (high-wire act) were integrated to create a stage of composite art.

Their promotion strategies also were extraordinary. They decided to distribute their promotion images through YouTube to reach a global audience from a long-term perspective. They quietly approached the youth around the world through new media such as SNS, which led to a point where young audiences around the world requested for K-pop performances. The same strategy was applied to Psy who is also popular in the world.

6.3 The Left Hand Law of Creativity

In the previous sections, we made the rules T1, T2, and T3 on the axis of time; S1, S2, and S3 on the axis of space; and F1, F2, and F3 on the axis of field. Asking questions on these axes is like moving by means of a navigator. You need to make it a habit to ask questions to improve your creativity in the process. To make "creating" a habit, repeated behaviors are to be involved, and the best method in this respect is "commendation". People tend to repeat a certain behavior when commended on it. This is because of dopamine that functions in the brain's reward system. The more you are commended, the more dopamine is generated, stimulating the nerves of pleasure. It makes you repeat the process to be commended and to gain pleasure from it.

The points that we have tackled are a mere theory. Merely understanding a theory is different from making it part of a habit. For example, let us think of a person good at reading music. Even if he works very hard to read music well, it does not guarantee an excellent performance. To make an outstanding performance a habit, you need to practice repeatedly up to the point of mastering your craft. To this end, this section introduces *The Left Hand law of Creativity* (Fig. 6.7).

The Left Hand law of Creativity begins with spreading your fingers as in the figure. This is similar to Fleming's left hand law which you learned at school. The three fingers represent a certain axis in a three-dimensional area with the middle finger representing the axis of time, the index finger representing the axis of space,

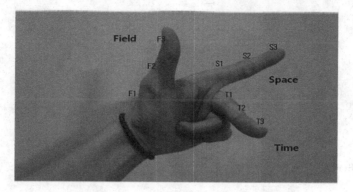

Fig. 6.7 Left hand law of creativity

and the thumb representing the axis of field. When facing a certain problem, spread your fingers as illustrated in the figure, and think of the three-dimensional left hand law of creativity. Think of questions you may ask on the axis of time, on the axis of space, and on the axis of field.

Coincidentally, our fingers have three knuckles. You may match each knuckle to one rule. For instance, you may assign each knuckle of the middle finger the axes of time (T1, T2, and T3) in the order of palm to finger tip. Likewise, assign the three knuckles of the index finger the axes of field (S1, S2, and S3). Assign the three knuckles of the thumb the three rules of field (F1, F2, and F3). When touching each knuckle on the axis, think of the corresponding rule.

Whenever you encounter a problem, spread your fingers in application of the left hand law of creativity, and apply rules in the order of each axis. By applying the rules in a TSF three-dimensional world, you can solve various problems.

6.4 Circuits of Brain Cells

Recognizing and judging things is a process in the brain (Fig. 6.8). We learn new things from the point of birth to death. How would the brain remember such knowledge? How would a new thought be generated in the brain? How would our behaviors be decided and made?

There are about 100 billion nerve cells (neuron) in a human brain. Nerve cells in a brain are called neurons. These nerve cells consist of cell bodies, dendrites, axons, and synapses. A cell body is the primary body of a cell, which contains a nucleus (Fig. 6.9).

A dendrite receives signals from other nerve cells. An axon is a sort of communication cable to deliver signals to other nerve cells. A synapse at the edge of an axon delivers the signals to other brain cells. The signals received through the dendrite are sent to the cell body. Signals from many dendrites are united in a cell

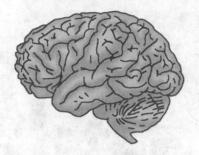

Fig. 6.8 Brain

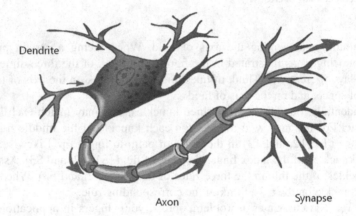

Fig. 6.9 Nerve cell

body, passed through the dendrite then the synapse, and then moved to another cell. The signals running in nerve cells are electrical.

According to findings, the thoughts and memories of humans are formed in the cooperative circuits among nerve cells. Nerve cells connect to one another rather than function independently.

This is similar to the way electronic circuits in a semiconductor chip work. The electronic elements in a chip cannot function independently. Various elements form a small circuit with each circuit performing a respective function that unites with other functions to work in one integrated chip (Figs. 6.10, 6.11).

6.5 Habits are a Product of Neural Circuits

Nerve cells are connected not merely because they are close to each other but because they have a common objective. Even if they are far from each other, nerve cells may be connected when necessary as electronic elements in a circuit are.

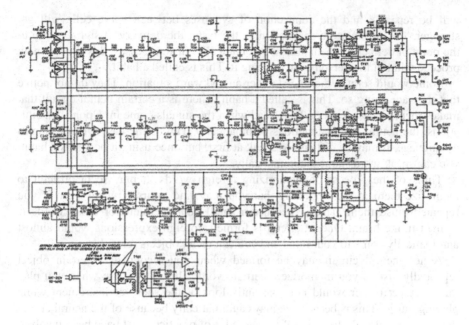

Fig. 6.10 Electric circuit

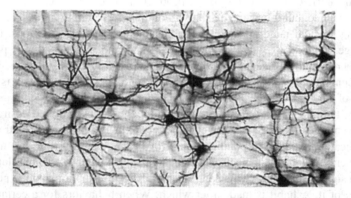

Fig. 6.11 Neural circuit

How is the image of a tiger remembered in a brain? The area that handles memories in a brain consists of a number of nerve cells. Not all of these cells are connected. Some are connected for form a neural circuit, which holds memories.

To hold a new memory of a tiger, a few nerve cells are connected to form a network, and this becomes a neural circuit that remembers a tiger. You need to note here that this sort of neural circuit is formed when there are multiple external stimuli.

This is why we need to repeat an act many times to memorize a new word. Suppose you often see a tiger. The process of delivering signals as stated earlier will be repeated and the connection of synapses between nerve cells will get stronger. The stronger the connection is, the faster the delivery of signals is up to the point of delivering them almost automatically. In some cases, this process may proceed without the person even noticing it. This is called a habit.

Some would scratch their head under an awkward situation. They do not notice that they are doing so. This is called a habit. There is a certain neural circuit that makes them place a hand on their head when the signals come in. Upon such situations, the neural circuit functions automatically. This is the case when it comes to language. It may be difficult to speak at first, but once using it becomes a habit, you can speak words almost automatically.

Think of the process of memorizing foreign words. It may seem difficult to memorize them at the beginning, but you may find yourself using them if you repeatedly use them. This is because a small neural circuit for those words is formed in the brain. Once a circuit is formed, foreign expressions would almost automatically come to your mind and you can recall objects related to a word.

A new neural circuit may be formed when you recognize a certain object repeatedly just as you memorize a certain word. This is the case when you play music. A performer would not constantly look at the score or instrument while playing music. This is because he plays automatically because of the neural circuit that has been already formed. Of course, a lot of practice must have been involved to form this sort of circuit.

The mannerism of scratching one's head in an awkward situation might have begun by chance, but it became a habit without the person noticing it.

Playing musical instruments or scratching the head involve motions. Learning a language, in contrast, involves mental habits. Some may see things positively while others see them negatively. Mental habits make such a difference. It is said that certain paradigms are involved when people see things, which is another example of a mental habit.

The Pygmalion Effect means that when someone believes in and desires something, he comes to achieve it in the end. Believing in and desiring something to be done indicates that a corresponding circuit of brain cells is forming in the brain. Once the circuit is formed, electronic signals flow to it when there are related stimuli. Such signals make the corresponding neural circuit function and the person act in pursuit of it. A habit is made after which. When behaviors for a certain desire are repeated, the possibility of achieving it increases accordingly.

These new neural circuits are not always formed in the brain, however. It is quite difficult to overcome a weak point if the person is born with it. For instance, a person born left-handed may have to put in a tremendous amount of effort to become right-handed. Some may be born ineffective at sports. These persons have not genetically inherited athletic abilities and they have to work harder to become good at sports though it is not impossible. To overcome an inborn weakness, more effort than usual is necessary.

You may find very weak but multiple connections among nerve cells if you observe a brain photograph of a child. There may not be many neural circuits in a child's brain yet, but they may develop as the child grows, learns a language, and recognizes objects around him. Circuits formed in childhood may remain longer since they are used more often. That is why memories of childhood events are vivid and habits made in childhood years hardly change. The saying "Old habits die hard" is correct in terms of brain science. This explains why the elderly appear stubborn. As a person grows old, the connection of frequently used neural circuits becomes stronger while that of hardly used ones deteriorate. As a result, the neural circuits are likely to reject new things.

Existing circuits may disappear if you do not use them. This is the case when you forget the name of an old friend or when you find it difficult to think of words when you do not use a language.

6.6 Watching TV Upside Down

Some say that we are born with personalities. This is partially true. If someone lives a certain way without spending extra effort to change his ways, his life would totally depend on his inherited disposition. In this type of person's brain, neural circuits that perform only physical instincts may develop. You may be able to manipulate neural circuits, however, if you try hard through learning. In other words, required habits may overcome inherited elements since habits are made in combination of inborn elements and effort over time.

This is the case when it comes to talent. Some may be born with talents, but even such talents may not be known if people do not train. It may be easy to develop an inborn talent if an individual recognizes it earlier and tries hard to develop it. Even if you are born without a specific skill, your effort may result in its development.

Although a certain talent is not inherited, you may develop it better than a person born with it if you devote a lot of effort and consistent interest. Daniel Letivin calls this "the principle of ten thousand hours" in his book *This is Your Brain on Music.* It says that if you invest as much as ten thousand hours in a certain practice, you can get a fruitful outcome.

Some say that you can do nothing about creativity since creativity is an inborn trait and cannot be learned. True, it is something innate. As discussed earlier, however, something innate can be changed as well. Both habit and creativity can be changed if you do not give up and if you exert effort consistently.

I watch TV upside down as I mentioned in Chap. 3. I stand my TV set upside down while watching it. I started doing this six years ago to test how much I could change myself. I watched TV upside down at least 10 minutes a day. At first, it seemed very weird, but now I am used to it.

There is nothing I cannot do in the world after I learned how to see things upside down. There is neither a fixed idea nor an impossibility. Trees and buildings are upside down, and people walk upside down. Everything looks different as I see things upside down. Even familiar people look totally different. I can even read subtitles at the bottom of the screen while looking them upside down.

The habit of watching TV upside down is a result of making a new neural circuit in the brain. I proved to myself that a neural circuit could be formed through artificial means. Now I am practicing using chopsticks with my left hand even though I am right-handed. I have a habit of putting my right foot first in the trouser when I put on a pair of trousers, but I am now trying to make it a habit to put in the left foot first.

Some ask me, "What is the benefit of watching TV upside down?" I reply, "Watching TV upside down is like calisthenics of the brain." If you do calisthenics, stretching your body in bed in the morning, your body will be relaxed and flexible. If you activate brain cells that are not often used, you will feel better as the brain becomes more flexible.

6.7 Make Yourself More Creative

As mentioned earlier, when a new circuit is formed in the brain, a new habit is formed and your personality also is gradually changed. Indeed, this event involves effort to change neural circuits and habits, but it is not impossible.

Habits may decide almost everything of the person. A person's personality depends on his habits which are related to language, behavior, knowledge, and character. Diligence is a habit. Sincerity is a habit. Honesty is a habit. Talking well is also a habit. When such a habit changes, the personality changes as well. As the personality changes, a life may change. Changing habits may lead to changing an entire life.

To make yourself change, seek to create new cell circuits in your brain. To do this, you have to repeat certain behaviors. Making a new cell circuit in the brain will change your habits. Making it a habit to ask questions to yourself when you are alone will help you produce a lot of fresh ideas. Practicing all these will help you tap your creativity.

The three-dimensional questions on time, space, and field are basic elements that can apply to anything. Anything in the world is in some way related to these three elements. Thus, the three-dimensional questions for general purposes are applicable to every problem.

Asking such three-dimensional questions is like moving through a three-dimensional navigation system, which will help a person break his adherence to reality. Once we move away from the reality of time, space, and field, it becomes easy to overcome fixed ideas and think in an integrated manner. Such a three-dimensional navigation provides us with a "frame of thinking" for asking ourselves. Expanding our thinking through the three navigations result to "problem-solving methods" with which we can handle any given problems.

Breaking away from reality does not mean putting the given problem out of mind. Be sure to have the given problem in mind when pursuing a TSF 3-D travel because your goal is to create fresh ideas while asking questions on a given issue.

Humans are likely to give up when they are ignorant of a specific principle. Once you understand a principle, apply it while working to achieve a certain objective. In the past, when a person caught a disease, he accepted it as a destiny and prayed for its cure. Today, however, people would try to have their illnesses healed.

The natives of Africa who are ignorant of the principles of car mechanisms, on the other hand, would perceive a car breaking down a matter of destiny and pray for it. Those who are ignorant of the principle regarding creativity may say that inventiveness is something innate. They regard creativity in terms of fate, give up enhancing it, and do nothing.

Now we understand the principle of creativity. Have confidence, repeat asking the three-dimensional questions, and work hard. You may then become a creative person with new neural circuits formed in your brain.

6.8 Summary

1. In the three-dimensional world the number of rules to ask questions is 27 (3 × 3 × 3).
2. A sense of humor is a result of applying T1, S3, and F3.
3. A chestnut peeling machine is a result of applying T2, S1, and F1 to existing products.
4. The success of K-pop, which is part of the Korean pop culture, is a result of the three-dimensional approach to the world.
5. The left hand law of creativity visualizes the method for three-dimensional creativity development.
6. The middle finger in the left hand law of creativity indicates T1, T2, and T3 in order.
7. The index finger in the left hand law of creativity indicates S1, S2, and S3 in order.
8. The thumb in the left hand law of creativity indicates F1, F2, and F3 in order.
9. Watching TV upside down is a sort of brain stretching.
10. Once you understand the basic principle of creativity, you can do away with the fixed idea that creativity is something innate and there is nothing you can do about it.

6.9 Exercise

1. Discuss what would change if you apply the TSF three-dimensional questions to the lessons in a classroom.
2. Discuss what would change if you apply the TSF questions to special transport companies such as Fedex or DHL.

3. Discuss what would change if you apply the TSF questions to developing a set of kitchen knives as a name brand.
4. Think of the nine rules that you can make by combining the thumb and index finger in the left hand law of creativity.
5. Think of the nine rules that you can make by combining the thumb and middle finger in the left hand law of creativity.
6. Discuss the role of synapses with regard to brain cells.
7. What parts of the brain function as the signal input and output ports when brain cells interact with one another?
8. Why would memories disappear if you did not use them for a long time?
9. Practice watching TV upside down.
10. Discuss innate factors and acquired ones with regard to talent. Discuss to what extent you may change personal abilities through acquired factors.

Bibliography

Lee, Kwang H. 2012. *Developing three dimensional creativity (Korean)*. Business Map Press.

Starko, Alane Jordan. 2010. *Creativity in the classroom*. London: Routledge press.

Kaufmann, James C., Sternberg, Robert J. (ed.). 2010. *Cambridge handbook of creativity*. Cambridge: Cambridge university press.

Lee, Kwang H. 2012. *Lecture notes on leadership and communication*. Graduate school of science journalism KAIST.

Lee, Kwang H. 2012. *Lecture notes on creativity*. GIFTED (Global Institute for Talented Education), KAIST.

Lee, Kwang H. 2013. Three dimensional creativity: three navigations to extend our thoughts. *Journal of advanced computational intelligence and intelligent informatics* 17(2):157–160

K. H. Lee, *Three Dimensional Creativity*, KAIST Research Series,
DOI: 10.1007/978-94-017-8804-5, © Springer Science+Business Media Dordrecht 2014

Printed in the United States
By Bookmasters